FRONTIERS IN ELECTRONIC TESTING

Volume 39

Series Editor:
Vishwani D. Agrawal
ECE, Auburn University, Auburn, Alabama, USA

More information about this series at http://www.springer.com/series/5994

Victor Champac • Jose Garcia Gervacio

Timing Performance of Nanometer Digital Circuits Under Process Variations

Springer

Victor Champac
INAOE
Tonantzintla, Puebla, Mexico

Jose Garcia Gervacio
Universidad Veracruzana
Xalapa, Veracruz, Mexico

ISSN 0929-1296
Frontiers in Electronic Testing
ISBN 978-3-030-09239-9 ISBN 978-3-319-75465-9 (eBook)
https://doi.org/10.1007/978-3-319-75465-9

Printed on acid-free paper

This Springer imprint is published by the registered company Springer International Publishing AG part of Springer Nature.
The registered company address is: Gewerbestrasse 11, 6330 Cham, Switzerland

To my son Francisco, and our families.

Preface

Integrated circuits are present in a diversity of electronic applications around us. We can find them in a wide spectrum of consumer applications like smartphones, iPods, video games, mobile computing, and printers, among others. Electronics applications are present in servers and networking and also in critical applications like medical, space, avionics, and cars. Electronics products are expected to play an important role in new market applications such as the Internet of things (IoT). Possibilities of these applications range from smart-home applications, wearable devices, and industrial automation to smart cities among others in this growing market. The success of electronic applications has relied on a continuous scaling of semiconductor technologies, which has allowed an increase of performance and complexity at constrained cost. On the other hand, the difference in the performance characteristics of a given circuit fabricated at a large number of units has become significant at scaled semiconductor technologies in the nanometer regime. The differences in performance are caused by different conditions the die suffers during manufacturing, which is known as process variations. These have an important impact on the yield and quality of the circuits and finally on the profitability of the semiconductor companies.

Process variations have been traditionally a concern in analog circuits, but to a lesser extent in digital circuits. However, they are playing a major role in digital circuits as the semiconductor technologies have scaled in the range of tens of nanometers. Traditional corner-design methodologies to deal with process parameter variations in digital circuits have become less efficient with technology scaling as the impact of the intra-die variations has increased. Conventional design methods to deal with process variations incur high-cost penalties, which grows with technology scaling. Because of this, the modern digital design requires considering the statistical nature of the process variations and how they impact circuit performance. Advanced semiconductor technology nodes have brought the need for engineers with a good understanding of process variations and their impact on circuit behavior. This book pursues to provide the readers a comprehensive understanding of the nature of the variations that occur during fabrication, how they are modeled and their interactions, and how they impact the performance of modern and future integrated circuits. Emphasis is dedicated to providing digital designers with an understanding of how the process variations affect the behavior of the circuits they are working to be able to improve their performance. Understanding process variations is quite important to satisfy the demand for high-performance requirements of actual integrated circuits, which is getting even more important with the continuous scaling of semiconductor technologies.

Important features of this book are:

- Engineering students and engineer practitioners in industry will find a book with topics oriented to the digital design of integrated circuit under process variations.
- Theoretical fundamental expressions followed by natural circuit level analysis are used for a better and smooth understanding of the behavior of logic cells and circuits under process variations. But the reader can also obtain some important knowledge of the behavior of digital circuits under process variations even without going deeper into mathematics.
- The book covers all the required subjects for understanding its main goal.
- Logical sequence treatment of learning topics. A step-by-step sequence methodology, going from logic gates to logic paths to circuit level, is followed to explain the behavior of process variations at the different levels.
- Important *design hints* as a result of the analysis presented in the book are highlighted.
- This book contains some guided examples useful to practice and to reinforce the theoretical concepts presented to the reader.
- This book provides useful appendixes to the readers.

The project of this book was conceived more than one decade ago. An important question the authors raised themselves was how we would like someone to teach us about this topic? This was further enriched by the student comments over the years, and constant discussions with designers in industry have been quite important for this project. Another question was how to efficiently help engineer practitioners, who have a limited amount of time to understand about the nature of process variations and how they impact their designs to be able to make decisions for improving their designs. Moreover, some mathematical background is included in the book for a good understanding, but a reader can obtain some essential insights about circuit performance under process variations even skipping the mathematics. This book will bring to engineer students and engineer practitioners skills for improving circuit performance leading to electronic products with higher quality and yield. Even more, better design productivity and improved time to market can be achieved. The readers may foresee new possibilities and new directions of research for improving the performance of actual and future digital integrated circuits.

A brief description of the chapters of this book follows.

Chapter 1: Introduction This chapter provides the general framework for the book. An overview of the semiconductor technology evolution through years is presented, and the main driving forces for technology scaling are presented. A design flow for modern digital integrated circuits to obtain a final designed circuit working on required performance within product specifications is described, and the role of process variations to achieve the previous goals is outlined. The corner-design methodology, widely used in industry to deal with process variations, and how statistical analysis overcomes its limitations are discussed.

Chapter 2: Mathematical Fundamentals This chapter gives an introduction to topics of probability and statistical theory in the scope of the goal of this book. Discrete and continuous random variables along with uniform and normal distributions are explained. Concepts of covariance and correlation and the sum of normal random variables are described. Additionally, a brief explanation of the theorem of Taylor is described.

Chapter 3: Process Variations In this chapter, a brief overview of the main steps to fabricate circuits in CMOS technology is given. Afterward, the most important sources of process variations due to non-idealities of the manufacturing process are identified and described. The behavior and modeling of the different types of process variations (i.e., inter-die and intra-die variations) is discussed. Models to account for spatial correlation are presented.

Chapter 4: Gate Delay Under Process Variations This chapter studies the impact of the process variations on the statistical performance of logic cells. The impact of process variation on the timing performance of an inverter cell is analyzed with detail, and a methodology to compute its statistical gate delay is described. The statistical treatment is extended to a 2-Nand gate and generalized to other types of logic gates. Several key design issues affecting the standard deviation of the gate delay are illustrated.

Chapter 5: Path Delay Under Process Variations This chapter studies the statistical timing performance of a logic path composed of several gates under process variations. The methodology to compute the statistical delay of a two-inverter chain is illustrated, and then it is extended to a general logic path. Pure random and correlated variations are considered. Several key design issues affecting the path delay are illustrated, such as the impact of spatial correlation, the impact of relative delay sensitivities between gates, sizing a logic path, and logic depth. The main advantages of statistical-based design over corner-based design are analyzed.

Chapter 6: Circuit Analysis Under Process Variations This chapter studies the impact of process variations on the delay performance of some illustrative circuits of general interest. The circuits are analyzed using the methodologies described in the previous chapters and using Monte Carlo analysis with SPICE. A 1-Bit Full Adder, 4-Bit Carry Ripple Adder, and a benchmark ISCAS circuit are considered. Some measures that the designer may take to improve these designs are discussed. Afterward, the performance of a 6T-SRAM cell memory under process variations is analyzed along with a high-sigma analysis.

Chapter 7: FinFET Design Issues and Process Variations Impact This chapter describes FinFET devices and some important fabrication steps used in the manufacturing of FinFET circuits. Circuit design issues unique to FinFET technology are discussed, and a step-by-step procedure to create the layout of an inverter cell is presented. The main sources of process variations in FinFET technology are discussed, and their impact on the performance of the logic cells is discussed. The computing of the statistical delay of an inverter gate based on FinFET technology is presented.

Appendix A This appendix illustrates the approximation of the mean and standard deviation of a function composed of several normal random variables by using a Taylor series expansion of first order.

Appendix B This appendix describes the methodology to simulate a logic gate under process parameter variations (pure and correlated random variations) using SPICE electrical simulator.

Appendix C This appendix describes a methodology to account for the spatial correlation of parameters between gates of a logic path that are separated by a certain distance in SPICE electrical simulator.

The material of this book has been enhanced in the lecture courses related to digital VLSI design given through the years at the National Institute for Astrophysics, Optics, and Electronics (INAOE). Feedback from the students has been valuable. The authors would like to thank many persons for contributing in some way to the book. Victor Avendaño collaborated in the Section of *Nanometer Design* and Hector Villacorta in the Section of *Corner-based Nanometer Design* of Chapter one. Special thanks to Andres Gomez, Freddy Forero and Javier Mesalles for their contribution to the book, they made a lot of SPICE electrical simulations and worked on other issues.

Tonantzintla, Puebla, Mexico — Victor Champac
Xalapa, Veracruz, Mexico — Jose Garcia Gervacio

Contents

About the Authors

Victor Champac received the Electrical Engineering Degree in 1987 from the Autonomous University of Nuevo Leon, Mexico. He received the Ph.D. degree in 1993 from the Polytechnic University of Catalonia (UPC), Spain. From 1988 to 1993 he was Associate Professor at the Electronic Engineering Department of the UPC. In 1993 he joined the National Institute for Astrophysics, Optics and Electronics (INAOE) in Mexico where he is Titular Professor. He made sabbatical in 2001–2002 at Motorola and in 2010–2011 at the University of California (UCSD). Dr. Champac was a co-founder of the Test Technology Technical Council-Latin America of IEEE Computer Society. He was co-General Chair of the 2nd, 9th, 13th and 16th IEEE Latin-American Test Workshop (now Latin-American Test Symposium). He has also served as Guest Editor of the Journal of Electronics Testing, Springer. He is member of the Editorial Board of Journal of Electronic Testing (JETTA). He is a Senior member of the IEEE and the National Research System (SNI) in Mexico. He has served as program and organizing committee member of several international conferences. Dr. Champac received the best paper award of the IEEE Latin American Test Workshop in 2008. He has published more than 120 papers in international journals and conferences. His research lines include: circuit design under process variations, aging reliable circuit design, defect modeling in leading technologies, and development of new test strategies.

Jose Garcia Gervacio received the Electronics Engineering Degree in 1999 from the Technological Institute of Celaya, Mexico. He received the Master and Ph.D. degree in 2003 and 2009, respectively, from the National Institute for Astrophysics, Optics, and Electronics (INAOE), Mexico. From 2010 to 2011 he was an assistant professor at INAOE. From 2011 to 2013 he occupied a postdoctoral position in the Research center in Micro and Nanotechnology (MICRONA) of the University Veracruzana. Since 2013 he is Full-time Professor of the University Veracruzana in Xalapa, Mexico. He has served as program and organizing committee member of the 4th Biannual European—Latin American Summer School on Design, Test, and Reliability (BELAS-2015). He served as organizing committee member of the 16th IEEE Latin American Test Symposium (LATS-2015). His research interest is focused on the design and testing of VLSI digital circuits.

About the Authors

Acronyms

ADC	Analog-Digital Converter
ASIC	Application-specific integrated circuit
BEOL	Back-End-Of-Line
BIST	Built-in Self-Test
BJT	Bipolar Junction Transistor
CAD	Computer-aided design
CD	Critical dimension
CDF	Cumulative distribution function
CMOS	Complementary Metal-Oxide Semiconductor
CMP	Chemical mechanical polishing
D2D	Inter-die variations
DAC	Digital-Analog Converter
DRC	Design Rule Checking
DSP	Digital signal processing
ERC	Electric Rule Checking
FEOL	Front-End-Of-Line
FER	Fin LER
FinFET	Fin field-effect transistor
FSM	Finite-State Machine
GER	Gate LER
IC	Integrated circuit
JTAG	Joint Test Action Group
LELE	Litho-Etch Litho-Etch
LER	Line edge roughness
LWR	Line Width Roughness
MOL	Middle-Of-Line
MOSFET	Metal-Oxide Semiconductor Field-Effect Transistor
nMOS	N-type MOS transistor
OPC	Optical proximity correction
OPE	Optical proximity effect
PDF	Probability density function

PDN	Pull-Down Network
PMF	Probability mass function
pMOS	P-type MOS transistor
PUN	Pull-Up Network
PVT	Process-Voltage-Temperature variations
RCA	Ripple Carry Adder
RDF	Random dopant fluctuation
RET	Resolution enhancement technique
RTL	Register-Transfer Logic
RV	Random variable
SADP	Self-aligned double patterning
SoC	System-on-Chip
SOI	Silicon-On-Insulator
SRAM	Static Random Access Memory
SSTA	Statistical Static Timing Analysis
STA	Static Timing Analysis
t_{PHL}	High-to-low propagation time
t_{PLH}	Low-to-high propagation time
VHDL	VLSI hardware description language
VLSI	Very Large Scale Integration
WFV	Work function variation
WID	Intra-Die variations

Chapter 1
Introduction

Abstract This chapter provides the general framework for the book. An overview of the semiconductor technology evolution through years is presented, and the main driving forces for technology scaling are presented. A design flow for modern digital integrated circuits to obtain a final designed circuit working on required performance within product specifications is described, and the role of process variations to achieve the previous goals is outlined. The corner design methodology, widely used in industry to deal with process variations, and how statistical analysis overcomes its limitations are discussed.

1.1 Semiconductor Technology

Electronic applications have been increasing through time. Vacuum tubes allowed that electronics being used in applications traditionally covered by mechanical devices. Bipolar transistor was an important advance in semiconductor technology. Then, MOSFET devices allowed to increase the complexity of the functions covered by electronics. More recently, multi-gate FinFET devices claim the continued improvement of the advantages of semiconductor devices. The advent of semiconductor devices made possible a widespread of electronic applications. The work of Bardeen and Brattain lead to demonstrate the point-contact transistor [1–3], and shortly after in 1948, W. Shockley invented the Bipolar Junction Transistor (BJT) [1, 2, 4, 5]. The basic structure of the BJT consists of two junctions located close to each other (Fig. 1.1a). The behavior of this device is based on a diffusion mechanism. Thermally excited electrons are injected from the emitter into the base due to the forward bias of the base-emitter junction [6, 7]. The carriers diffuse through the base from the emitter side to the collector side due to differences in the concentrations. These electrons are swept to the collector by the reverse-biased base-collector junction. The BJT allowed the implementation of basic digital gates, medium complex digital functions, memories of limited capacity, and even some microprocessors. However, they presented disadvantages of high-power consumption, large area, high voltages, and also scalability problems [8].

V. Champac, J. G. Gervacio, *Timing Performance of Nanometer Digital Circuits Under Process Variations*, Frontiers in Electronic Testing 39,
https://doi.org/10.1007/978-3-319-75465-9_1

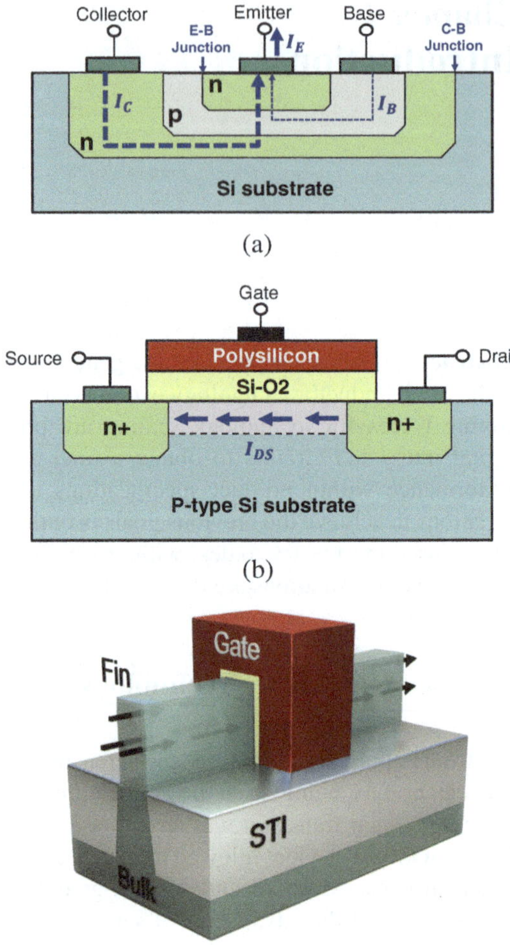

Fig. 1.1 Semiconductor devices. (**a**) BJT transistor. (**b**) MOSFET transistor. (**c**) FinFET bulk transistor

The realization of the insulated-gate field-effect transistor (FET) was a key-driven event in the electronic revolution. A patent of the insulated-gate field-effect transistor (FET) was issued as earlier as in 1929 by Julius Lilienfeld [4, 9]. At that time this device was not possible to manufacture. The first successfully insulated-gate field-effect transistor was built at Bell Labs by M.M. Atalla and D. Kahng in 1959 [2, 10, 11]. The main structure of the MOSFET (metal-oxide semiconductor field-effect transistor) consists of an insulation material placed between a gate electrode and a semiconductor (Fig. 1.1b). The behavior of this device is based on a drift mechanism [6]. The current in the surface of the semiconductor is modulated by a transverse electrical field applied to the gate electrode. MOSFET devices offered better scaling properties, which allowed to obtain smaller transistors with performance improvement.

MOSFET devices made possible the realization of more complex circuits at lower-power consumption. Furthermore, the power consumption was further significantly reduced with the use of the Complementary MOS (CMOS) technology. The use of both nMOS and pMOS transistors to construct the Boolean functions in CMOS technology [12] allowed the implementation of complex microprocessors, high-density memories, digital signal processors (DSPs), and system-on-chip (SoC) design among others. Multi-gate devices have appeared more recently to continue with improvement in performance and complexity of electronic applications. The main advantages of multi-gate devices are the improved short-channel effects and reduced leakage [13–16]. Tri-gate FinFET transistor (e.g., FinFET transistor) is one of the multi-gate devices more feasible of being manufactured. In this structure [17, 18], both the gate oxide and the gate electrode are wrapped around a silicon fin (Fig. 1.1c). The electrical field at the gate electrode creates three channel inversions in the fin regions adjacent to the gate oxide. Thus, the transistor current flows along the side surfaces and top of the fin as shown in Fig. 1.1c. This technology is intended to allow the fabrication of even smaller circuits and microprocessors at reduced power consumption.

Interconnecting transistor devices construct logic functions J. Kilby in 1958 realized at Texas Instruments the first integrated circuit by putting more than one single electronic component on the same semiconductor chip [2, 19, 20]. Gold wires connected the electronic components. R. Noyce in 1960 fabricated at Fairchild Semiconductor the first working monolithic IC composed of four bipolar transistors and five resistors [2]. Aluminum metal strips made the connections between electronic components. C.T. Sah and Frank Wanlass in 1963 discovered that a logic function using p-channel and n-channel transistors consumed almost negligible power consumption in standby mode [2]. The use of both types of transistors was called CMOS technology. Commercial CMOS integrated circuits were made by RCA in 1969 [4, 21, 22]. Logic functions as Nor and Nand gates, 4-bit full adder, and decade counter among others were built.

Advances in semiconductor technologies have continued along the time leading to circuits with higher speed and higher integration density. Figure 1.2 illustrates the timeline of Intel processors since 1970 until 2010 (https://www.intel.la/content/www/xl/es/history/history-intel-chips-timeline-poster.html). It is observed that the clock speed and the transistor count have been increasing approximately at an exponential rate through the years. Intel 4004 processor appeared in 1971 presented a clock speed of 108 kHz and 2300 transistors, and the second generation of Intel core processors appeared in 2010 presented a clock speed of 3.8 GHz and 1.16 billion of transistors. These processors show the high complexity of modern integrated circuits, which are actually used in a wide spectrum of applications.

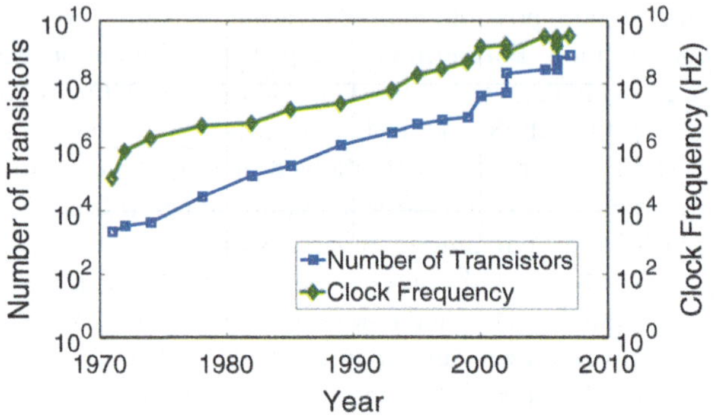

Fig. 1.2 Timeline of Intel processors (https://www.intel.la/content/www/xl/es/history/history-intel-chips-timeline-poster.html)

1.2 First Words of Process Variations on Semiconductor Technologies

The steps for manufacturing integrated circuit involve several chemical and physical processes, which take place on a silicon wafer [8, 23]. Deposition step places a desired material over the wafer, the photolithography step transfers a desired pattern to the wafer, and the etching process removes the undesired material. These three steps are repeated several times through the manufacturing process of an IC. Planarization using chemical mechanical polishing (CMP) and ion implantation are also two important steps among others. As the chemical and physical conditions vary from die-to-die, wafer-to-wafer, and lot-to-lot in manufacturing, the characteristic of the final devices and interconnections in the dies also vary [24]. Furthermore, the characteristics of the transistors and interconnects may fluctuate differently even in the same die. The fluctuations in the characteristics of the dies are known as *process variations*.

The photolithography process is an important source of process variations impacting the channel length and the channel width of the transistors. The chemical mechanical process during planarization of the surface of a material impacts significantly variations at the interconnects. Below 65 nm technologies, random dopant fluctuations turned into a relevant source of variations affecting the threshold voltage of the transistor [25–27]. As a consequence of the process variations, the speed of the circuits do not present a single value at the fabricated dies but a range of speed values. Figure 1.3 illustrates the delay distribution of a hypothetical set of fabricated dies of the same circuit where a normal delay distribution approximates their speeds. Most of the chip speeds are around the typical value; however, a significant number of chips take speed values slower/faster than the typical speed value.

Process variations have become more prominent with technology scaling [25, 26, 28]. Figure 1.4 shows the percentage of variations of some transistor and interconnect parameters. It is observed that the variation of the channel length (L_{eff}) of the

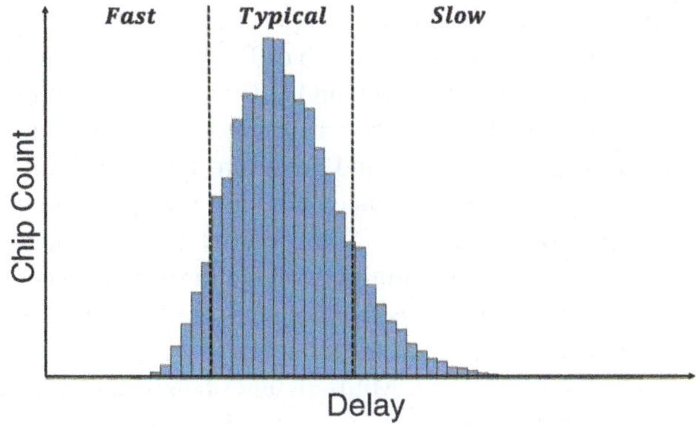

Fig. 1.3 Delay distribution of a logic path

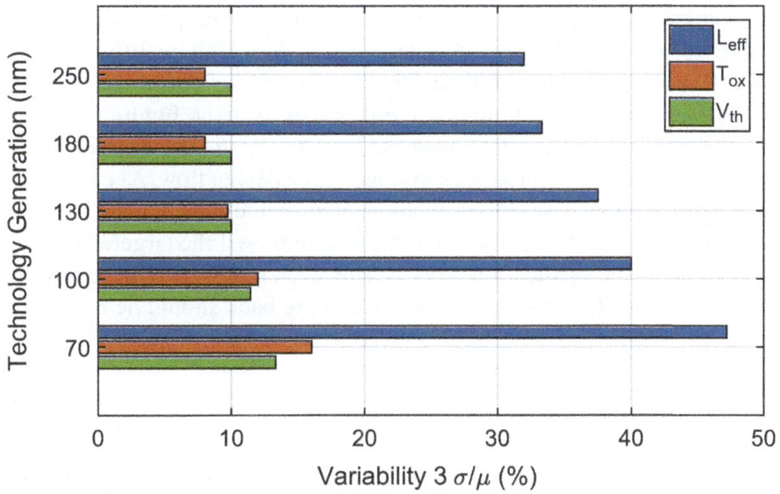

Fig. 1.4 Variations with technology scaling [26]

transistor increases considerably over technology scaling. The variations of other transistor parameters such as T_{ox} and V_{th} also increase with technology scaling.

1.3 Making Modern Digital Circuits

1.3.1 Nanometer Design

Modern digital designs are usually made by first setting a digital library known as standard cell library. The digital library is composed of a set of basic digital gates (e.g., inverter, Nand, Nor gates), buffers, complex cells, multiplexers, transmission gates, latches, flip-flops, I/O cells, interfaces, etc. The digital library may contain complex cells like memories. Logic cells with different driving capabilities to fulfill the desired requirements of speed, power, and area are included in the library. Cells with the higher driving capability (high-performance cells) are for a faster timing response, but they have higher power and area. On the other hand, cells with the lower driving capability (low-performance cells) are for a slower timing response, and they have lower power and area. Some digital libraries may contain more specialized cells constructed using multiple voltage threshold transistors (multi-V_{th} transistors) and multiple channel transistor lengths (multi-channel length) [29–31]. Multi-V_{th} digital libraries may be used for reducing power consumption [29, 32, 33] and in mixed-signal design [34, 35]. Multi-channel length digital libraries may be used to reduce leakage power [30, 35].

The design flow of VLSI circuits can present some differences by each design group at semiconductor companies. However, a design flow converges in general steps to achieve the overall goal after fabrication: to obtain a design working on required performance within product specification with high quality and the highest yield and the most important on-time. The time-to-market could be for most of the IC companies one of the most important milestones to cover. Figure 1.5 illustrates a typical top-down design flux of an integrated circuit [8, 36–38]. The specification of a given product must be verified at each stage of the design flow. At each abstraction level of the design flow, a set of verifications items should be defined. The designer may go back at each abstraction level of the design flow if the targeted requirements are not satisfied by the verification process. However, going back in the flow diagram increases product cost. Design changes when going back should be minimized.

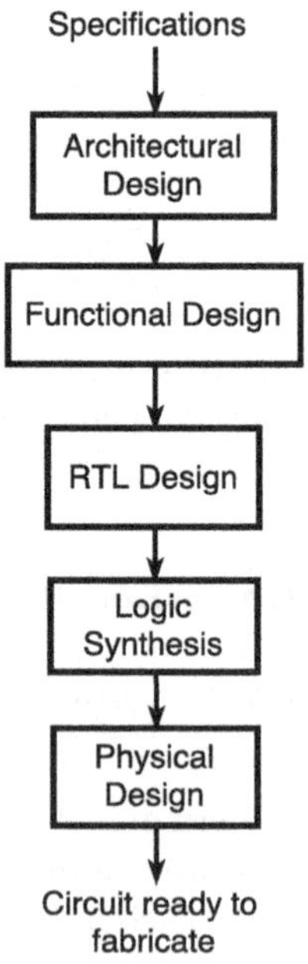

Fig. 1.5 Design flux of an IC

The chip starts with the *architectural design* stage of the circuit [8, 36–38]. At this level, the overall functionality of the system is defined by high-level block functions (e.g., processors, controllers, DSPs, memory, DACs/ADCs, interfaces, block functions of a general purpose logic, etc.) and the relationships between them. The behavior of the system is modeled and validated with tools like MATLAB or a high-level language like C++. The speed, area, and power consumption among others are estimated. Decisions whether a function is implemented in hardware or software are made at this level [37].

The *functional design* step [8, 36–38] translates the content of each one of the high-level block functions defined at the system design level into a behavioral description, which is usually based on a hardware description language (HDL) as VHDL or Verilog. The behavior of a unit is defined in terms of inputs, outputs, and timing without focusing on the specific hardware that will be used later for the unit [36]. This behavioral representation of a design gives flexibility and opportunity to the engineers for proposing solutions according to the desired goals of the specifications of the circuit.

A next step is to converge the previous behavioral description into a Register-Transfer Logic description (*RTL design step*) [8, 36–38]. The behavioral description of the functionality of the design is translated to Boolean functions. In this way, logic functional blocks are described to transfer information between storing registers. This description is also usually made using a hardware description language. An essential task is to verify that the functionality of the logic description complies with the intended overall chip functionality defined by the specifications of the design [39]. This verification is usually made using functional and formal verifications. Quality metrics like functional coverage and code coverage (e.g., FSM coverage, toggle coverage, etc.) are covered. The RTL version with a certain level of maturity is delivered to start the logic synthesis.

The *logic synthesis* translates the RTL code to a gate-level circuit [8, 36–38]. The functionality of the Boolean functions is mapped to logic cells of a digital library of a given technology satisfying performance requirements. The first deliverable in this stage is the gate-level netlist which can be released to the verification team. Functional verification can be run at zero delay just to ensure that nothing was broken during the translation from the behavioral code to gate level. In this step, the circuit design constraints such as clock system definitions and timing corner definitions among others are carefully verified. Typical IC designs need to be testable [40–42] and to comply with some test standards like IEEE 1149.1/.6 (JTAG) [43–46]. Some circuitry dedicated for the test must be included and built scan chains with the sequential elements in the design to test for stuck 0/1 faults at the fabricated dies [47]. Test insertion stage is at the end an important step in the flow since it must be included without impacting the overall performance of the circuit. Functional and test mode (i.e., BIST, scan, and JTAG) are verified to cover all the logic in the IC and achieve the higher test coverage as possible.

In the *physical design* stage [8, 36–38], each logic cell and their interconnections are translated to geometric entities (circuit layout) that will be used for fabrication of the chip. Based on a floor plan previously reviewed by architecture and design teams, place and route of the logic cells and interconnections that represent the design take place. Some verifications of the layout are made such as Design Rule Checking (DRC), Electrical Rule Checking (ERC), Layout Versus Schematic (LVS), and Antenna Rules [38]. Static timing analysis (STA) [48, 49] is made to get timing requirement in all paths in the design and for all required design corners (timing closure) [48, 49]. Typically design corners are specified according to process variability, a range of temperature, and a (+/−) percentage variation of nominal voltage (known as PVT corners). Timing violations in any of these PVT corners must be fixed before the final design is released. Some design groups make more effort to verify PVT corners before physical design. Normally, PVT analysis represents a bottleneck in the IC design flow.

After the physical design is successfully verified, a final GDSII file is delivered to manufacture the circuit.

1.3.2 Impact of Process Variation on Nanometer Design

A designer is interested to know the impact of the variations that the parameters suffer during fabrication on circuit performance specifications [25, 28]. Delay is an important performance specification that may take different values at the fabricated dies. Variations also occur for other performance specifications such as leakage, power consumption, noise margin, etc. Figure 1.6 illustrates the normalized frequency variation versus the normalized leakage current for the same circuit at different fabricated dies [50]. The delay and frequency are normalized to their minimum values. It is observed that both the delay and frequency do not have a single value but a range of values. The spread of the range of values of the frequency is 0.3×, and that of the leakage is 25× with 1× denoting the minimum value that the performance variable can take. The spread of leakage is more significant than that of frequency due to the exponential dependence of the subthreshold current on the threshold voltage, which takes different values due to process variations.

Process variations may also impact the noise margin tolerance of dynamic circuits [51, 52]. Figure 1.7 shows a circuit composed of a two-stage domino logic.

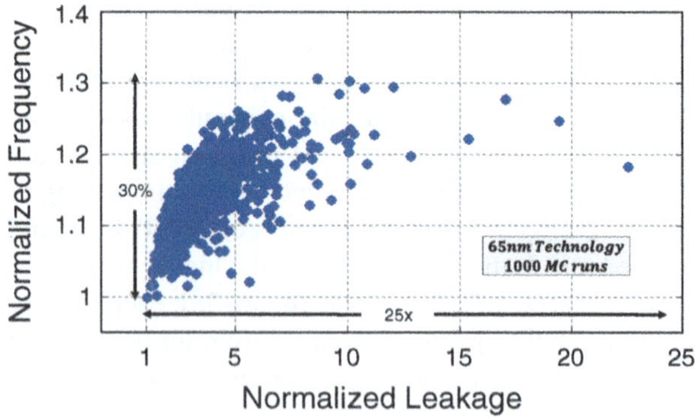

Fig. 1.6 Spread-out of delay versus leakage due to process variations

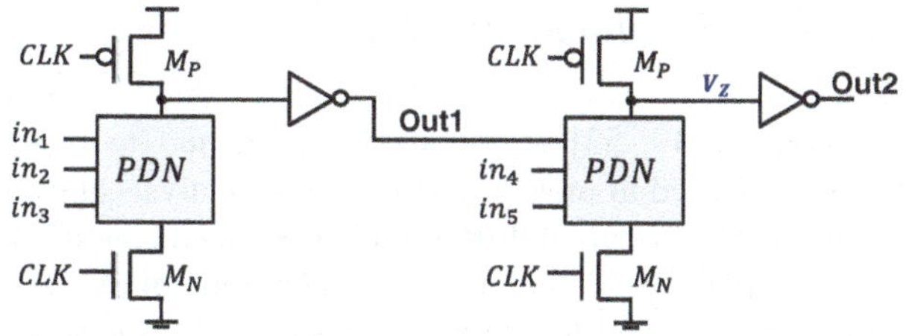

Fig. 1.7 A two-stage domino logic circuit

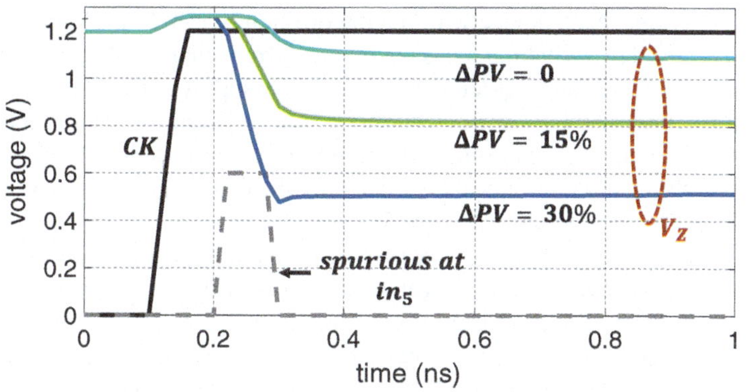

Fig. 1.8 Behavior of a two-stage domino logic under process variations

The operation of the dynamic logic relies on the precharge and evaluation phase. The outputs are precharged when the clock is at a low state (CK = 0). For the evaluation phase, the clock changes from low to high, and the domino logic evaluates according to the state of the circuit inputs (in_1 to in_5). Since the output of a domino logic may be in a high impedance state during the evaluation phase, this logic is susceptible to noise at the inputs. Process variations in the parameters of the evaluation transistor may affect the correct functionality of the cell. Let us assume that the precharge-evaluate stage of the second Domino cell (Fig. 1.7) is in high impedance state after the evaluation phase is completed. The logic states of the nodes are $in_4 = 1, in_5 = 0$, and $V_Z = 1$. The PDN network of the precharge-evaluate stage of the second cell may turn on leading to a wrong evaluation if a spurious pulse occurs at input in_5 and the threshold voltage (V_{th}) value of the evaluation transistor of this cell is lower than expected due to process variations. Figure 1.8 illustrates the behavior of the two-stage Domino circuit for three cases of variations of the parameters (W, L, T_{ox}, and V_{th}). The percentage of variation is defined with ΔPV. For ΔPV $= 0$ and ΔPV $= 15\%$, the high impedance node V_Z (Fig. 1.7) falls below V_{DD}, but its value is well above VDD/2. As a consequence, the two-stage domino logic presents correct logic functionality. However, for ΔPV $= 30\%$, the node V_Z falls well below VDD/2. Because of this, a wrong evaluation may occur.

1.3.3 Corner-Based Nanometer Design

Corner analysis [12, 26, 49, 53] is the standard method to deal with process variations of ICs widely used in industry. This type of analysis allows verifying the desired performance of the circuit at different PVT extreme conditions in which the circuit is expected to operate. PVT refers to a combination of process variation (P), power supply variation (V), and temperature variation (T). *Inter-die* and *intra-die* variations are two types of variations of concern to designers [24, 53, 54]. *Inter-die* variations affect equally the parameters of the devices within a die, while *intra-die* variations affect each device within a die differently [53, 54]. The power supply variation is caused by switching gates leading to current changes flowing through the resistance, inductance, and capacitance of the power supply network [25, 55, 56]. An important fraction of the power consumption due to circuit operation is converted to heat, which affects the operating temperature of the transistors [25, 56]. Besides, the working chip temperature is also affected by the external environment (e.g., in a car) where the circuits operate [25, 56, 57].

Figure 1.9 shows a parameter space for nMOS and pMOS transistors due to the process variations [12, 58, 59]. TT refers to typical nMOS-typical pMOS process parameter condition; SS and FF refer to slow nMOS-slow pMOS and fast nMOS-fast pMOS. SF and FS are known as mixed-mode corners and refer to slow nMOS-fast pMOS and fast nMOS-slow pMOS, respectively. The four corners of the box do not define a perfect square due to correlations [12]. As a consequence, the degree of slow (fast) nMOS (pMOS) transistor at SS (FS) corner is different than the degree of slow (fast) nMOS (pMOS) transistor at SF (FF) corner. For power supply voltage variations, the following conditions are usually considered: typical voltage (TV), high voltage (HV), and low voltage (LV). For temperature variations, the following conditions are usually considered: typical temperature (TT), high temperature (HT), and low temperature (LT).

A *design corner* is defined at a point (SS, SF, FF, FS, and TT) in a process parameter space (Fig. 1.9) with specific temperature and power supply voltage extreme conditions. Table 1.1 illustrates a possible set of PVT *design corners* for

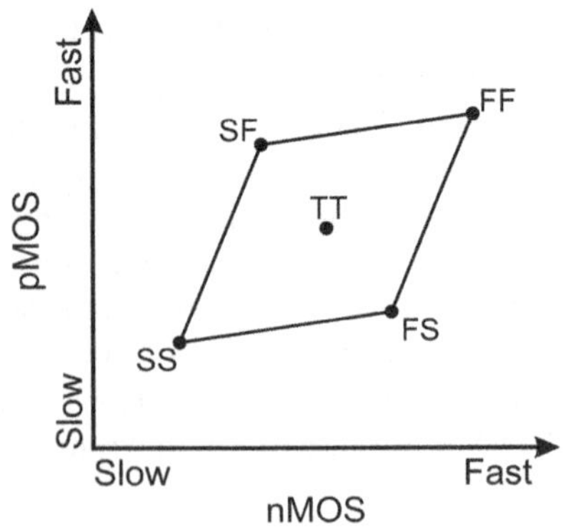

Fig. 1.9 Parameter space of transistor process variations [12]

Table 1.1 Process, Voltage, Temperature (PVT) design corners for V_{th}

Nmos V_{th} (V)	Pmos V_{th} (V)	VDD (V)	T (°C)	Design corner P V T
0.44	−0.42	1.2	80	TT TV TT
0.5	−0.48	1.09	125	SS LV HT
0.5	−0.48	1.09	−40	SS LV LT
0.34	−0.33	1.34	125	FF HV HT
0.34	−0.33	1.34	−40	FF HV LT

Table 1.2 Delay of a ten-inverter chain across PVT

	Maximum	TT TV TT	**SS LV HT**	SS LV LT	FF HV HT	FF HV LT	Status
Delay (ps)	44	39	**45**	42	36	32	Fail

Table 1.3 Delay of modified ten-inverter chain across PVT

	Maximum	TT TV TT	**SS LV HT**	SS LV LT	FF HV HT	FF HV LT	Status
Delay (ps)	44	37.5	**43.2**	41	34	30	Pass

the threshold voltage parameter (V_{th}) of a CMOS technology. Mixed-mode corners are not included.

PVT corner analysis allows that the IC designer finds worst-case circuit behavior such as the maximum and minimum values of a performance metric. Thus, the IC designer should simulate the circuit at all corners to ensure that the specifications are fulfilled at all the possible operating conditions. For example, let us consider a digital logic path composed of a ten-inverter chain with a maximum delay specification of 44 ps. Table 1.2 shows the path delay of the circuit across the corners given in Table 1.1. It is observed that the circuit fails at SS LV HT corner. A possible solution to fulfill the specification is to increase the size of the first inverter of the chain. Now, the path delay of the circuit fulfills the specification at all corners as shown in Table 1.3.

Newer technologies present important local variation not usually considered in corner analysis. The impact of these local variations can be considered by verifying the expected circuit performance doing a Monte Carlo simulation at each corner. The use of extensive Monte Carlo simulations represents a considerable cost on computing time.

1.4 Need of Statistical Circuit Design

Reducing overdesign is an important driving force for using statistical design [58, 60]. Fabricated circuits present a spread of the speed operation around the specific target nominal value due to process variations during manufacturing (Fig. 1.3). Those circuits with delay values located to the right of the target delay do not satisfy the specification.

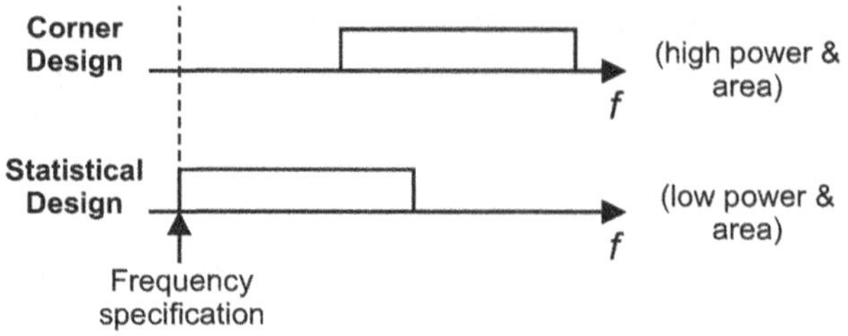

Fig. 1.10 Cost comparison of design methodologies [60]

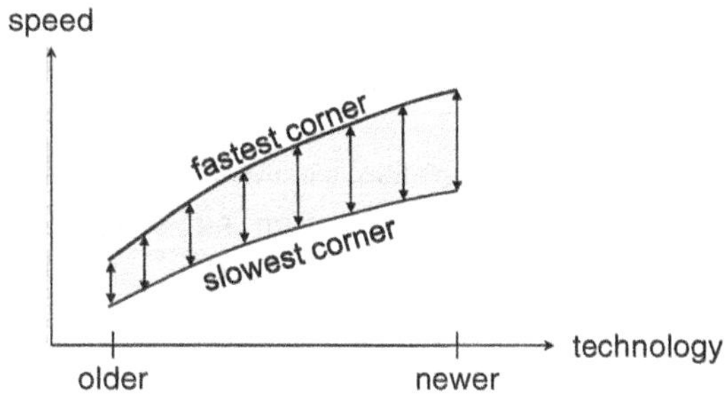

Fig. 1.11 Spread of circuit speed between the fastest and slowest corners

The corner-based design is a methodology widely used to deal with process variations [26, 49, 53]. Corner-based design verifies circuit functionality for pessimistic values of the technological parameters due to the process variation during fabrication. At the slowest corners, the parameters are assumed to take simultaneously those extreme values that reduce the current capability of all the transistors in the circuit, and as a consequence, the logic paths of a circuit also are assumed to have unrealistic worst-case delays. Under this assumption, transistors are sized up (e.g., cell gates of the library with a higher current capability are used) to fulfill the target delay specification of the logic paths of the circuit. Because of this, logic paths are sized up in excess (overdesigned) because all the transistors are assumed simultaneously weaker than they are in a fabricated circuit. Figure 1.10 presents an illustration of the cost incurred due to corner-based design [58–60]. Corner-based design leads to chips that are faster than their frequency specification. On the other hand, the statistical design allows that the chips fulfill the frequency specification more efficiently. Globally, chips designed with corner-based design present more power and area penalties than chips designed with statistical design. In Chap. 5, a detailed analysis of the cost incurred by worst-case-based design in comparison to statistical-based design is presented.

The cost of overdesign has increased with scaling of CMOS technologies [56, 58, 60–62]. Figure 1.11 illustrates the speed of a circuit designed at the fastest and slowest corners with technology evolution. The speed spread increases in newer technologies. In other words, the speed window between the fastest and the slowest corners increases with technology scaling. As a consequence, the cost of circuits

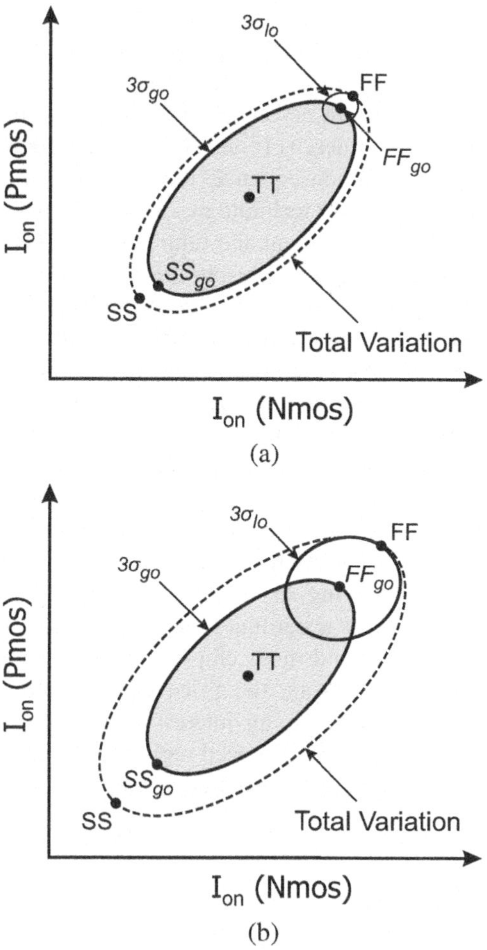

Fig. 1.12 Process variations with scaling. (**a**) Older technologies. (**b**) Newer technologies

designed using corner-based design becomes higher for newer technologies, which strongly motivates the use of statistical design in advance technology nodes.

Figure 1.12 shows the relative impact of global (σ_{go}) and local (σ_{lo}) variations in older and newer technologies [59, 63]. Global variations, also known as inter-die variations, present a radial dependence through the wafer and were successfully considered in the past using worst-case analysis [64, 65]. This was because the impact of local variations was significantly less severe than inter-die variations in older technologies. This is illustrated in Fig. 1.12a where the spread of the saturation current of nMOS and pMOS transistors due to local variations ($3\sigma_{\text{lo}}$) is much smaller than for global variations ($3\sigma_{\text{go}}$). The final corners (FF and SS) are close to the corners without local variations (FF_{go} and SS_{go}). In older technologies, corner-based analysis could guarantee a good yield of a circuit after fabrication. However, local variations due to the nature of their statistical behavior are pessimistically modeled by worst-case analysis [56, 64]. This fact aggravates in newer technologies, which present significant local variations. Figure 1.12b shows that the relative impact of intra-die variations ($3\sigma_{lo}$) with respect to global variations ($3\sigma_{go}$) increases significantly in newer technologies. Now, there is a significant difference between the final corners (FF and SS) and the corners without local variations (FF_{go} and SS_{go}). As a consequence, the cost incurred by corner-based design becomes even higher in newer technologies and shows the necessity of using statistical analysis in the design of current and future semiconductor technologies.

Underdesign and respin are other two important forces for proper modeling of variations in nanometer digital circuits [66]. Underdesign is opposite to overdesign. Circuits underdesigned may fail to fulfill performance specifications leading to yield loss [67, 68]. In the past, respins were mainly due to functional design errors, but in modern nanometer design, there is a growing concern to reduce/avoid respins due to process variations.

The fabricated chips present a spread of their performance characteristics such as speed and power consumption among others. A technique called speed-binning is a practice used by the manufacturers [69–73]. This technique consists in the application of input vectors to a chip at increasing frequencies until it fails, which allows knowing its maximum operating frequency. The chips are classified in different bins according to their speed (Fig. 1.3). Those chips with higher speed are sold higher and those chips with lower speed are sold cheaper. In AMD Opteron family processors, the price can vary for almost $200 to over $800 [69, 74]. Certainly, there is an interest of the vendors to move chips in the lower-speed region to the higher-speed region to increase chip revenue. Statistical-based design methodologies may be used to improve profit under process variations [75].

References

1. R. Reid, *The CHIP: How Two Americans Invented the Microchip and Launched a Revolution*, 2nd edn. (Random House Trade Paperbacks, New York, 2001)
2. F. Maloberti, A.C. Davos, *A Short Story of Circuits and Systems* (IEEE/Rivers Publishers, New York/Aalborg, 2016)
3. J. Bardeen, W.H. Brattain, Physical principles involved in transistor action. Bell Syst. Tech. J. **28**(2), 239–277 (1949). https://doi.org/10.1002/j.1538-7305.1949.tb03634.x
4. C. Piguet, History of low-power electronics, in *Low-Power CMOS Circuits Technology, Logic Design and CAD Tools*, ed. by C. Piguet. (Taylor and Francis Group, Boca Raton, 2006)
5. W. Shockley, A unipolar "field-effect" transistor. Proc. IRE **40**(11), 1365–1376 (1952). https://doi.org/10.1109/JRPROC.1952.273964
6. B. Razavi, *Fundamentals of Microelectronics* (Wiley India, Bangalore, 2009)
7. A.S. Sedra, K.C. Smith, *Microelectronic Circuits*, vol. 1 (Oxford University Press, New York, 1998)
8. R. Jacob Baker, *CMOS Circuit Design, Layout, and Simulation*, 3rd edn. (Wiley-IEEE Press, New York, 2010)
9. C. Kleint, Julius Edgar Lilienfeld: life and profession. Prog. Surf. Sci. **57**(4), 253–327 (1998). ISSN 0079-6816

10. D. Kahng, Electric field controlled semiconductor device, U. S. Patent No. 3,102,230 (Filed 31 May 31, 1960, issued August 27, 1963)
11. S. Chih-Tan, Evolution of the MOS transistor-from conception to VLSI. Proc. IEEE **76**(10), 1280–1326 (1988). https://doi.org/10.1109/5.16328
12. N.H.E. Weste, D.M. Harris, *CMOS VLSI Design A Circuit and Systems Perspective*, 4th edn. (Addison Wesley, Boston, 2011)
13. C. Hu, New CMOS devices and compact modeling, in *2007 International Workshop on Physics of Semiconductor Devices*, Mumbai, 2007, pp. 5–7. https://doi.org/10.1109/IWPSD.2007.4472441
14. T. Cui, Q. Xie, Y. Wang, S. Nazarian, M. Pedram, 7 nm FinFET standard cell layout characterization and power density prediction in near- and super-threshold voltage regimes, in *International Green Computing Conference*, Dallas, TX, 2014, pp. 1–7. https://doi.org/10.1109/IGCC.2014.7039170
15. L. Chang, K.J. Yang, Y.-C. Yeo, Y.-K. Choi, T.-J. King, C. Hu, Reduction of direct-tunneling gate leakage current in double-gate and ultra-thin body MOSFETs, in *International Electron Devices Meeting. Technical Digest*, Washington, DC, 2001, pp. 5.2.1–5.2.4. https://doi.org/10.1109/IEDM.2001.979428
16. B. Yu et al., FinFET scaling to 10 nm gate length, in *Digest. International Electron Devices Meeting*, San Francisco, CA, 2002, pp. 251–254. https://doi.org/10.1109/IEDM.2002.1175825
17. S.K. Hadia, R.R. Patel, Y.P. Kosta, FinFET architecture analysis and fabrication mechanism. Int. J. Comput. Sci. Issues **8**(5), 235–240 (2011)
18. P. Mishra, A. Muttreja, N.K. Jha, FinFET circuit design, in *Nanoelectronic Circuit Design*, ed. by N. Jha, D. Chen (Springer, New York, 2011)
19. J.S. Kilby, The integrated circuit's early history. Proc. IEEE **88**(1), 109–111 (2000). https://doi.org/10.1109/5.811607
20. In Memoriam - Jack Kilby (1923–2005) Inventor of the integrated circuit. IEEE Signal Process. Mag. **22**(5), 6–7 (2005). https://doi.org/10.1109/MSP.2005.1511813
21. M.J. Riezenman, Wanlass's CMOS circuit. IEEE Spectr. **28**(5), 44 (1991). https://doi.org/10.1109/6.83438
22. F.M. Wanlass, Low stand-by power complementary field effect circuitry, US Patent 3,356,858, Google Patents, 1967
23. H. Kaeslin, *Digital Integrated Circuit Design: From VLSI Architectures to CMOS Fabrication*, 1st edn. (Cambridge University Press, New York, 2008)
24. M. Dietrich, J. Haase, *Process Variations and Probabilistic Integrated Circuit Design* (Springer, New York, 2012)
25. M. Wirnshofer, *Sources of Variation* (Springer Netherlands, Dordrecht, 2013), pp. 5–14
26. C. Forzan, D. Pandini, Statistical static timing analysis: a survey. Integr. VLSI J. **42**(3), 409–435 (2009). Special Section on DCIS2006
27. S.S. Chung, The variability issues in small scale trigate cmos devices: random dopant and trap induced fluctuations, in *Proceedings of the 20th IEEE International Symposium on the Physical and Failure Analysis of Integrated Circuits (IPFA)*, July 2013, pp. 173–176
28. A. Asenov, Statistical device variability and its impact on design, in *2008 14th IEEE International Symposium on Asynchronous Circuits and Systems*, April 2008, pp. xv–xvi
29. J. Kao, A.P. Chandrakasan, Dual-threshold voltage techniques for low-power digital circuits. IEEE J. Solid State Circuits **35**, 1009–1018 (2000)
30. N. Sirisantana, L. Wei, K. Roy, High-performance low-power CMOS circuits using multiple channel length and multiple oxide thickness, in *International Conference on Computer Design*, 2000, pp. 227–232
31. B.P. Wong, A. Mittal, Y. Cao, G. Starr, *Nano-CMOS Circuit and Physical Design* (Wiley-Interscience, New York, 2005)
32. V. Melikyan, E. Babayan, A. Melikyan, D. Babayan, P. Petrosyan, E. Mkrtchyan, Clock gating and multi-VTH low power design methods based on 32/28 nm ORCA processor, in *2015 IEEE East-West Design & Test Symposium (EWDTS)*, Batumi, 2015, pp. 1–4. https://doi.org/10.1109/EWDTS.2015.7493159

33. V. Melikyan, T. Hakhverdyan, S. Manukyan, A. Gevorgyan, D. Babayan, Low power Open-RISC processor with power gating, multi-VTH and multi-voltage techniques, in *2016 IEEE East-West Design & Test Symposium (EWDTS)*, Yerevan, 2016, pp. 1–4. https://doi.org/10.1109/EWDTS.2016.7807678
34. S.A. Tawfik, V. Kursun, Low power and high speed multi threshold voltage interface circuits. IEEE Trans. Very Large Scale Integr. VLSI Syst. **17**, 638–645 (2009)
35. B. Amelifard, F. Fallah, M. Pedram, Low-power fanout optimization using multi threshold voltages and multi channel lengths. IEEE Trans. Comput. Aided Des. Integr. Circuits Syst. **28**(4), 478–489 (2009). https://doi.org/10.1109/TCAD.2009.2013992
36. N. Sherwani, *Algorithms for VLSI Physical Design Automation* (Kluwer Academic, New York, 1999)
37. H. Veendrick, *Nanometer CMOS ICs From Basics to ASICs* (Springer, Cham, 2008)
38. A.B. Kahng, J. Lienig, I.L. Markov, J. Hu, *VLSI Physical Design: From Graph Partitioning to Timing Closure* (Springer, New York, 2011)
39. B. Wile, J.C. Goss, W. Roesner, *Comprehensive Functional Verification* (Morgan Kaufmann, San Francisco, 2005)
40. N.M. Nayeem, J.E. Rice, A simple approach for designing online testable reversible circuits, in *Proceedings of 2011 IEEE Pacific Rim Conference on Communications, Computers and Signal Processing*, Victoria, BC, 2011, pp. 85–90. https://doi.org/10.1109/PACRIM.2011.6032872
41. P.K. Lala, An introduction to logic circuit testing. Synth. Lect. Digit. Circ. Syst. **3**(1), 1–100 (2008)
42. M. Abramovici, M.A. Breuer, A.D. Friedman, Design for testability, in *Digital Systems Testing and Testable Design*, vol. 1 (Wiley-IEEE Press, New York, 1990), pp. 343–419. https://doi.org/10.1109/9780470544389.ch9
43. A.T. Dahbura, M.U. Uyar, C.W. Yau, An optimal test sequence for the JTAG/IEEE P1149.1 test access port controller, in *Test Conference, 1989. Proceedings. Meeting the Tests of Time, International*, Washington, DC, 1989, pp. 55–62. https://doi.org/10.1109/TEST.1989.82277
44. P. Fleming, Semiconductor perspective on test standards, in *International Test Conference 1988 Proceeding New Frontiers in Testing*, Washington, DC, 1988, pp. 197–198. https://doi.org/10.1109/TEST.1988.207799
45. C.J. Sherratt, ICL's first development using IEEE 1149.1 (JTAG), in *IEE Colloquium on Application and Development of the Boundary-Scan Standard*, London, 1990, pp. 2/1–2/3
46. IEEE Standard for Test Access Port and Boundary-Scan Architecture, in *IEEE Std 1149.1-2013 (Revision of IEEE Std 1149.1-2001)*, May 13 2013, pp. 1–444. https://doi.org/10.1109/IEEESTD.2013.6515989
47. H. Bhatnagar, *Advanced ASIC Chip Synthesis* (Kluwer Academic, New York, 2002)
48. J. Bhasker, R. Chadha, *Static Timing Analysis for Nanometer Designs: A Practical Approach*, 1st edn. (Springer, New York, 2009)
49. S. Kilts, Static timing analysis, in *Advanced FPGA Design: Architecture, Implementation, and Optimization*, vol. 1 (Wiley-IEEE, New York, 2007), pp. 269–278. https://doi.org/10.1002/9780470127896.ch18
50. W. Wang, V. Reddy, V. Balakrishnan, S. Krishnan, Y. Cao, Statistical prediction of circuit aging under process variations, in *Solid State Circuits Technologies*, ed. by J.W. Swart (InTech, Croatia, 2010). https://doi.org/10.5772/6875
51. K. Yelamarthi, C.I.H. Chen, Timing optimization and noise tolerance for dynamic CMOS susceptible to process variations. IEEE Trans. Semicond. Manuf. **25**(2), 255–265 (2012). https://doi.org/10.1109/TSM.2012.2185961
52. M. Alioto, G. Palumbo, M. Pennisi, Understanding the effect of process variations on the delay of static and domino logic. IEEE Trans. Very Large Scale Integr. VLSI Syst. **18**(5), 697–710 (2010). https://doi.org/10.1109/TVLSI.2009.2015455
53. D. Blaauw, K. Chopra, A. Srivastava, L. Scheffer. Statistical timing analysis: from basic principles to state of the art. IEEE Trans. Comput. Aided Des. Integr. Circuits Syst. **27**(4), 589–607 (2008)

54. A. Chandrakasan, W.J. Bowhill, F. Fox, *Design of High-Performance Microprocessor Circuits* (Wiley, New York, 2000)
55. Q.K. Zhu, *Power Distribution Network Design for VLSI* (Wiley-Interscience, New York, 2014)
56. M. Orshansky, S. Nassif, D. Boning, *Design for Manufacturability and Statistical Design: A Constructive Approach*, 1st edn. (Springer, New York, 2010)
57. S. Kalra, Effect of temperature dependence on performance of digital CMOS circuit technologies, in *International Conference on Signal Processing and Communication (ICSC)*, 2013, pp. 392–395
58. M. Onabajo, J. Silva-Martinez, *Analog Circuit Design for Process Variation-Resilient System-on-a-Chip* (Springer, New York, 2012)
59. J. Ma, W. Yang, L. Yang, L. Wu, Z. Liu, On the modeling of process variations and its applications for circuit performance designs, in *2012 IEEE 11th International Conference on Solid-State and Integrated Circuit Technology*, Xi'an, 2012, pp. 1–4. https://doi.org/10.1109/ICSICT.2012.6467638
60. X. Li, J. Le, L.T. Pileggi, Statistical performance modeling and optimization. Found Trends Electron. Des. Autom. **1**(4), 331–480 (2007). http://dx.doi.org/10.1561/1000000008
61. B. Beacham, P. Hua, C. Lacy, M. Lynch, D. Toffolon, Mixed-signal IP design challenges in 28-nm process and beyond, Synopsys White Paper, July 2012
62. Y. Cao, L.T. Clark, Mapping statistical process variations toward circuit performance variability: an analytical modeling approach. IEEE Trans. Comput. Aided Des. Integr. Circuits Syst. **26**(10), 1866–1873 (2007)
63. R. Radojcic, D. Perry, M. Nakamoto, Design for manufacturability for fabless manufacturers. IEEE Solid-State Circuits Mag. **1**(3), 24–33 (2009)
64. A. Agarwal, D. Blaauw, V. Zolotov, Statistical timing analysis for intra-die process variations with spatial correlation, in *IEEE International Conference on Computer Aided Design (ICCAD)*, November 2003, pp. 900–907
65. B. Nikolic et al., Technology variability from a design perspective. IEEE Trans. Circuits Syst. Regul. Pap. **58**(9), 1996–2009 (2011). https://doi.org/10.1109/TCSI.2011.2165389
66. A. Gupta, DeepChipİ, http://deepchip.com/items/0573-03.html. June 2017
67. T. McConaghy, P. Drennan, *Variation-Aware Custom IC Design: Improving PVT and Monte Carlo Analysis for Design Performance and Parametric Yieldİ* (Solido Design Automation, Inc., San Jose) (2011)
68. M. Li, G. Huang, X. Wu et al., A yield-enhanced global optimization methodology for analog circuit based on extreme value theory. Sci. China Inf. Sci. **59**, 082401 (2016). https://doi.org/10.1007/s11432-015-0471-4
69. A. Das, S. Ozdemir, G. Memik, J. Zambreno, A. Choudhary, Microarchitectures for managing chip revenues under process variations. IEEE Comput. Archit. Lett. **7**(1), 5–8 (2008)
70. A. Datta, S. Bhunia, J. H. Choi, S. Mukhopadhyay, K. Roy, Profit aware circuit design under process variations considering speed binning. IEEE Trans. Very Large Scale Integr. VLSI Syst. **16**(7), 806–815 (2008). https://doi.org/10.1109/TVLSI.2008.2000364
71. L.Y.Z. Lin, C.H.P. Wen, Speed binning with high-quality structural patterns from functional timing analysis (FTA), in *2016 21st Asia and South Pacific Design Automation Conference (ASP-DAC)*, Macau, 2016, pp. 238–243. https://doi.org/10.1109/ASPDAC.2016.7428017
72. Q. Shi, M. Tehranipoor, X. Wang, L. Winemberg, On-chip sensor selection for effective speed-binning, in *2014 IEEE 57th International Midwest Symposium on Circuits and Systems (MWSCAS)*, College Station, TX, 2014, pp. 1073–1076. https://doi.org/10.1109/MWSCAS.2014.6908604
73. P. Das, S.K. Gupta, Extending pre-silicon delay models for post-silicon tasks: validation, diagnosis, delay testing, and speed binning, in *2013 IEEE 31st VLSI Test Symposium (VTS)*, Berkeley, CA, 2013, pp. 1–6. https://doi.org/10.1109/VTS.2013.6548901
74. AMD, AMD Processor Pricing, http://www.amd.com/pricing. May 2006
75. A. Datta, S. Bhunia, J.H. Choi, S. Mukhopadhyay, K. Roy, Speed binning aware design methodology to improve profit under parameter variations, in *ASP-DAC '06 Proceedings of the 2006 Asia and South Pacific Design Automation Conference*, 2006, pp. 712–717

Chapter 2
Mathematical Fundamentals

Abstract This chapter provides the basic mathematical fundamentals that are used throughout the book. These concepts allow a more precise understanding of the impact of process variations on the performance modern digital integrated circuits. First, the chapter gives an introduction to the probability and statistical theory. Then, discrete and continuous random variables are described. Next, the chapter describes the mean and variance, which enable the characterization of random variable functions. Some major distributions of random variables such as the uniform and normal distributions are described. Covariance and correlation that are two important concepts in the context of this book are explained. Statistical SUM operator for independent and correlated random variables is described. Besides, a brief explanation of the theorem of Taylor is given.

2.1 Basic Definitions

The probability and statistical theory are commonly used in many areas of knowledge where it is required to process, to analyze, and to understand an amount of information (dataset). In general, managing data of statistical class can be performed using many techniques and tools; it depends on the problem to solve. Some important definitions related to probability and statistical theory are given below.

2.1.1 Definitions

Random process: It is a process or experiment whose result cannot be predicted with certainty [1, 2]. For instance, the rolling of a die.

Outcome: It is the result of a process or experiment [1, 3]. For instance, the number or symbol labeled face upward on the die after rolling a die.

V. Champac, J. G. Gervacio, *Timing Performance of Nanometer Digital Circuits Under Process Variations*, Frontiers in Electronic Testing 39,
https://doi.org/10.1007/978-3-319-75465-9_2

Variable: It is a number or quantity that can be measured or counted. Its value can vary depending on conditions and over time [1, 4].

Random variable: A kind of variable whose values are defined by the possible outcomes of a random process [1, 4].

2.2 Random Variables

As defined above, a random variable is the one whose values are determined by a random process. According to the nature of this process, random variables can be classified into *discrete* or *continuous* as described below.

2.2.1 *Discrete Random Variables*

A discrete random variable X is the result of a random experiment that produces a countable number of outcomes [1, 4]. The discrete random variable X only can take values from the finite set $x_1, x_2, \ldots, x_n$ of possible outcomes of the random experiment. Figure 2.1 in the left side illustrates a typical random experiment of rolling a die that leads to a discrete random variable. This random experiment has six possible outcomes $(1, 2, \ldots, 6)$. Another experiment is the toss of a coin, which has only two possible outcomes (head or tail).

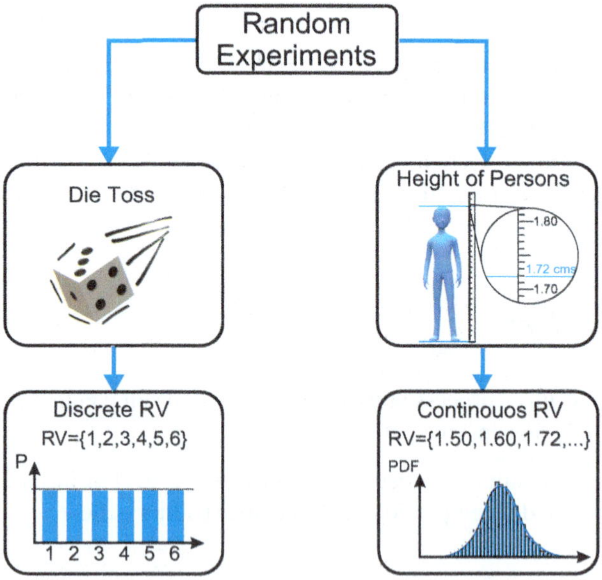

Fig. 2.1 Random experiments leading to discrete and continuous random variables

A discrete random variable is modeled by its probability mass function and/or its cumulative distribution function [1, 2, 4–10]. They are described below.

2.2.1.1 Probability Mass Function

The *probability mass function* (PMF) is a function that contains the information of all the possible discrete values the variable X can take and their probabilities [1, 4]. The probability corresponding to each value depends on the nature of the random experiment. For instance, in the random experiment of rolling an unbiased die, each outcome has the same probability. However, if the die is biased, some outcomes take a higher probability. The PMF gives the probability that the discrete random variable takes a given value as a result of a random experiment [1, 4]:

$$p(x_i) = P[X = x_i] \tag{2.1}$$

The previous equation gives the probability ($p(x_i)$) that the discrete random variable X takes a value x_i [1, 4]. In a valid PMF, the probability assigned to each outcome of the random variable must be equal or greater than zero, and the sum of the probabilities of all the outcomes must be the unit.

2.2.1.2 Cumulative Distribution Function

The *cumulative distribution function* (CDF) is a function that gives the probability that the discrete random variable takes values lower or equal than a specific value. The CDF is given by [1, 4]

$$F(x) = P[X \leqslant x] = \sum_{x_i \leqslant x} p(x_i) \tag{2.2}$$

The previous equation gives the probability $F(x)$ that the discrete random variable X takes a value equal than or lower than x [1, 4]. The probability $F(x)$ is obtained by summing up the probabilities of each one of the discrete outcomes ($p(x_i)$) equal than or lower than x. The CDF $F(x)$ may take values between 0 and 1, and the sum of all the probabilities $p(x_i)$ is equal to 1.

Example 1 A classical random experiment is the rolling of a die. Assuming the die is biased, each outcome has a different associated probability. Let X be a random variable that represents the outcomes of the die with six faces labeled 1 to 6. The values that the random variable can take are in function of the outcomes. The probability mass function $p(x_i)$ of each outcome x_i of the random variable X is given in the next table.

x_i	1	2	3	4	5	6
$p(x_i)$	0.1	0.25	0.15	0.15	0.25	0.1

Figure 2.2a displays the probability mass function of the die, represented by the random variable X. There are three probability values for the six possible outcomes of the die; this is because the die is biased. Figure 2.2b shows the cumulative distribution function of X. It is observed that the value of the distribution function $F(x)$ increases in each outcome (x_i) of the die and the increment in $F(x)$ corresponds to the associated probability $(p(x_i))$ of the outcome. $F(x)$ goes from 0 to 1 and is constant between successive outcomes; the maximum value is the sum of all the probabilities $p(x_i)$.

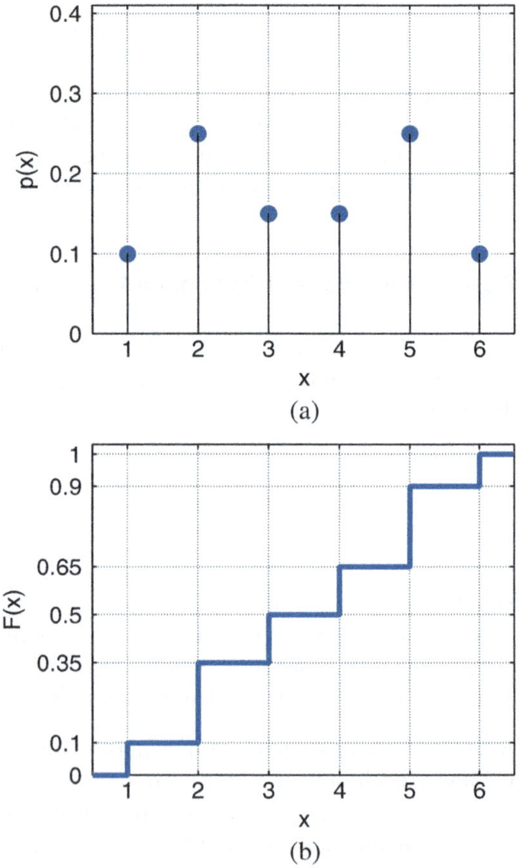

Fig. 2.2 Functions of a discrete random variable X. (**a**) Probability mass function (PMF). (**b**) Cumulative distribution function (CDF)

2.2.2 *Continuous Random Variables*

A continuous random variable X is the result of a random experiment that produces an infinite and non-countable number of outcomes [1, 4]. The continuous random variable X can take values from the infinite possible outcomes of the random experiment. Figure 2.1 in the right side shows the random experiment of measuring the height of persons in a population that leads to a continuous random variable. A person could have a height of 1.7 m or 1.71,1.72, 1.73, and 1.8. Also, we could have smaller fractions for measuring the height of a person like 1.7105 and 1.7106, and we could continue further having even smaller fractions of the height (i.e., there is an unlimited number of possibilities). Another experiment producing a continuous random variable is the measurement of the water's temperature, where again there is an unlimited number of possible outcomes. Other examples that can result in continuous variables include weight, mass, density, volume, pressure, length, etc.

A continuous random variable is modeled by its *probability density function* and/or its *cumulative distribution function* [1, 2, 4–10]. They are described below.

2.2.2.1 Probability Density Function

The *probability density function* (PDF) is a function that contains the information of the probabilities of the continuous values that the random variable X can take [6, 7]. The probability of continuous random variable X to take a value in the range between a and b is given by [6, 7],

$$P[a \leqslant X \leqslant b] = \int_a^b f(x)dx \tag{2.3}$$

The probability is obtained by the integration of the probability density function $f(x)$ in the range of interest defined between a and b. The probability is one if the PDF is evaluated over the entire space. It must be noted that the PDF cannot be defined for a single outcome but a set of outcomes taking values in a given range.

2.2.2.2 Cumulative Distribution Function

The *cumulative distribution function* (CDF) is a function that gives the probability that a continuous random variable takes a value less than or equal to a specific value, and it is expressed as [6, 7]

$$F(x) = P[X \leqslant x] = \int_{-\infty}^{x} f(x)dx \tag{2.4}$$

The previous equation gives the probability $F(x)$ that the continuous random variable X takes a value less than or equal to x. The probability $F(x)$ is obtained by the integration of the PDF from $-\infty$ to a value x. The CDF $F(x)$ takes values between 0 and 1.

Example 2 Let X be a random variable that represents the temperature of water in a pool. The probability density function which describes the likelihood of values that the random variable can take is given by $f(x) = 1/5$ in the range of $25 \leqslant x \leqslant 30$, otherwise $f(x) = 0$.

Using (2.3), the probability that a measure of the water temperature falls between 27 and 28 °C is as follows:

$$P[27 \leqslant X \leqslant 28] = \int_{27}^{28} \frac{1}{5} dx = 0.2 \tag{2.5}$$

Using (2.4), the probability that a measure of the water temperature is less than or equal to 28 °C is as follows:

$$F(x) = P[X \leqslant 28] = \int_{25}^{28} \frac{1}{5} dx = 0.6 \tag{2.6}$$

Figure 2.3a illustrates the probability density function (PDF) of the water temperature in the pool. The probability that a measure of the water temperature lies between the interval $[a, b]$ is equal to the area under the curve over that interval. For instance, the probability that a measure of the water temperature falls between 27 and 28 °C is $P[27 \leqslant X \leqslant 28] = 0.2$. Figure 2.3b illustrates the cumulative distribution function (CDF) of the water temperature in the pool. For instance, the probability that a measure of the water temperature is less than or equal to 28 °C is $P[X \leqslant 28] = 0.6$.

2.3 Characteristics of Random Variables

In this section, the mean and variance of a random variable are described.

2.3.1 Mean

The mean (first moment) of a random variable can be understood as the average value the variable takes after repeating the random experiment a very large number of times[1, 4–6, 10, 11]. The mean of X is usually denoted by μ_X or by $E[X]$, which is named the expected value of the random variable. The mean for the discrete and continuous cases is defined below.

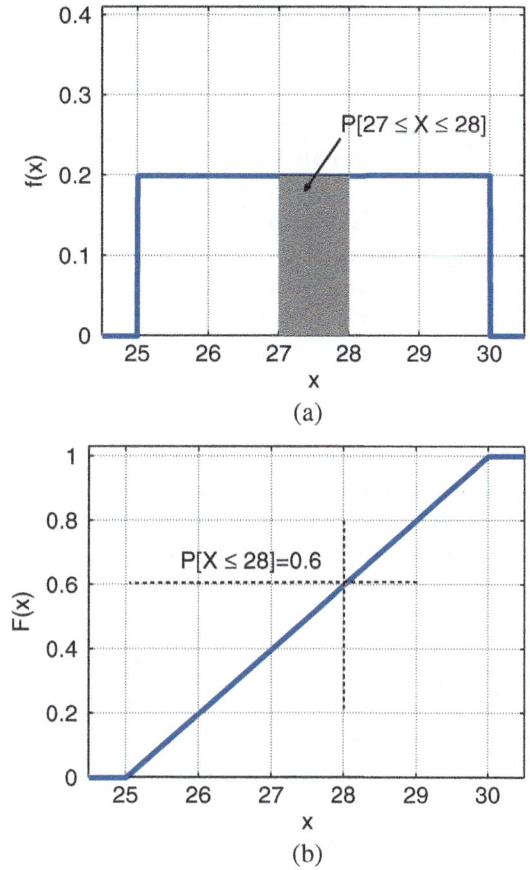

Fig. 2.3 (**a**) Probability density function (PDF) and (**b**) cumulative distribution function (CDF) of a continuous random variable X. (**a**) The shaded area in the PDF represents the probability that random variable X lies between the interval [27, 28]. (**b**) The cross-mark on the rising line indicates the probability that the random variable X is less than or equal to 28

2.3.1.1 Discrete Variables

The mean value of a discrete random variable X is obtained by summing up the products resulting from the multiplication of each value the variable takes (outcome) with the probability of that value to occur, and it is expressed as [1, 4]

$$\mu_X = E[X] = x_1 p(x_1) + x_2 p(x_2) + \cdots + x_n p(x_n)$$

$$\mu_X = E[X] = \sum_{i=1}^{n} x_i p(x_i) \tag{2.7}$$

In the previous expression, we are giving a degree of importance (a weight- $p(x_i)$) to each of the possible outcomes (x_i) of the random experiment. For instance, the mean of the dataset of the random variable X given in Example 1 is obtained using (2.7) as follows:

$\mu_X = E[X] = (1 * 0.1) + (2 * 0.25) + (3 * 0.15) + (4 * 0.15) + (5 * 0.25) + (6 * 0.1)$

$\mu_X = 3.5$

2.3.1.2 Continuous Variables

The mean value of a continuous random variable X is obtained by the integration of the continuous values the variable takes, represented by x, multiplied by the probability density function of the variable ($f(x)$) in a range bounded by a and b, and it is expressed as [1, 4]

$$\mu_X = E[X] = \int_a^b x f(x) dx \tag{2.8}$$

where $a < b$ and both can take values in the interval $(-\infty, \infty)$.

For instance, using (2.8), the mean of the random variable X given in Example 2 is obtained. The values this variable can take are defined by the probability density function $f(x) = 1/5$ in the range of $25 \leqslant x \leqslant 30$, otherwise $f(x) = 0$. The resulting mean is

$$\mu_X = \int_{25}^{30} x \frac{1}{5} dx$$
$$\mu_X = 27.5$$

2.3.2 *Variance*

The mean value alone gives important information about the random variables, but it is also needed to have information about the spreading of the values the variables take. We could have two random variables with the same mean but different spreading of the outcomes. The variance (second moment) is an important characteristic of a random variable that provides information about how far or close are the outcomes from the expected value [1, 4–6, 10, 11]. That is to say, the variance measures the spread of the outcomes from the mean value. A small variance denotes that most of the values the variable takes are near to the mean value, and a large variance indicates that the values are farther from the mean value.

The variance for the discrete and continuous case is described below.

2.3.2.1 Discrete Variables

The variance of a discrete random variable is obtained by summing up the squaring of each difference between the value the variable takes (outcome) and the mean multiplied by the probability of the outcome, and it is expressed as [1, 4]

$$\sigma_X^2 = \text{VAR}[X] = \sum_{i=1}^{n} (x_i - \mu_X)^2 p(x_i) \tag{2.9}$$

In the previous expression, $(x_i - \mu_X)$ evaluates how far or close is an outcome from the mean, and the squared difference is weighted by the probability of the outcome $p(x_i)$. For instance, using (2.9), the variance of the dataset of the random variable X given in Example 1 is obtained as follows:

$$\sigma_X^2 = \text{VAR}[X] = (1 - 3.5)^2 * 0.1 + (2 - 3.5)^2 * 0.25 + (3 - 3.5)^2 * 0.15$$
$$+ (4 - 3.5)^2 * 0.15 + (5 - 3.5)^2 * 0.25 + (6 - 3.5)^2 * 0.1$$
$$\sigma_X^2 = 2.45$$

2.3.2.2 Continuous Variables

The variance of a continuous random variable is obtained by the integration in the range of the squaring of the difference between the continuous values the variable takes x (outcomes) and the mean multiplied by the probability density function of the variable, where the integration is bounded by the interval $[a, b]$. The variance of the continuous random variable is given by [1, 4]

$$\sigma_X^2 = \text{VAR}[X] = \int_a^b (x - \mu_X)^2 f(x) dx \tag{2.10}$$

where $a < b$ and both can take values from $-\infty$ to $+\infty$.

For instance, using (2.10), the variance of the random variable X given in Example 2 is obtained. The values this variable can take are defined by the probability density function $f(x) = 1/5$ in the range of $25 \leqslant x \leqslant 30$, otherwise $f(x) = 0$. The variance result is,

$$\sigma_X^2 = \int_{25}^{30} (x - \mu_X)^2 \frac{1}{5} dx$$
$$\sigma_X^2 = 2.083$$

Standard Deviation The standard deviation also indicates the spread of values that the random variable takes. The standard deviation is the square root of the variance as given by [1, 4]

$$\sigma_X = \text{STD}[X] = \sqrt{\text{VAR}[X]} \tag{2.11}$$

The standard deviation is usually more convenient to use than the variance because it is easier to interpret as its unit is the same than the random variable.

2.4 Distributions of Random Variables

In probability and statistics theory, many distribution functions may describe the behavior of random variables [1–10, 12]. Two primary distribution functions widely used to model continuous random variables are the uniform distribution and normal distribution.

2.4.1 Continuous Uniform Distribution

A continuous random variable is said to follow a uniform distribution when its possible outcomes take a constant probability [3, 8]. Figure 2.4a illustrates the shape of the probability density function (PDF) of a uniform-distributed continuous random variable. The PDF of a random variable X uniformly distributed in the interval $[a, b]$ is given by [3, 8]

$$f(x) = \frac{1}{b-a} \quad \texttt{for } a \leqslant x \leqslant b \tag{2.12}$$

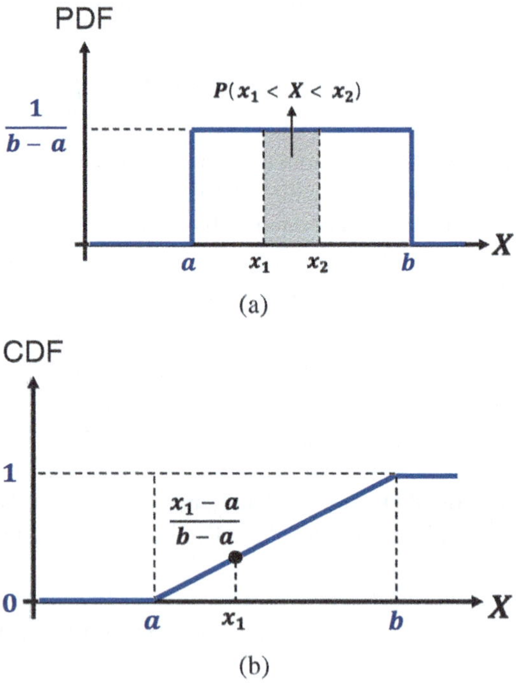

Fig. 2.4 (**a**) Probability density function (PDF) and (**b**) cumulative distribution function (CDF) of a uniform-distributed continuous random variable

As Fig. 2.4a shows, the PDF is zero out of the interval $[a, b]$, as those outcomes are not within the space of the random variable. The probability that X takes a value in the interval defined by x_1 and x_2 corresponds to the area of the shaded region, which is the integral of the PDF in the desired interval.

By applying the first moment (mean) and the second moment (variance), the following expressions are obtained:

$$E[X] = \tfrac{a+b}{2} \quad \mathrm{VAR}[X] = \tfrac{(b-a)^2}{12} \tag{2.13}$$

The mean value of a uniform distribution function is located exactly in the middle of the interval of the possible outcomes $[a, b]$. The variance, which measures how much the data are spread from the mean value, depends on the distance between the interval bounds.

The shape of the cumulative distribution function (CDF) of a uniform-distributed continuous random variable is shown in Fig. 2.4b. It is obtained by the integration of the PDF from the lower limit (a) to the given value of interest (i.e., $X = x_1$). As X increases within the interval $[a, b]$, the CDF linearly increases from 0 to 1. Mathematically, the CDF is defined by [3, 8]

$$F(x) = \frac{x-a}{b-a} \quad \texttt{for } a \leqslant x < b \tag{2.14}$$

2.4.2 Continuous Normal Distribution

One of the most important distributions to model the behavior of random variables is the well-known *normal distribution* (also called Gaussian distribution). A continuous random variable is said to follow a normal distribution if the shape of its PDF is as shown in Fig. 2.5a, and it is given by the following equation [3, 8]:

$$f(x) = \frac{1}{\sigma\sqrt{2\pi}} \exp^{-\frac{(x-\mu)^2}{2\sigma^2}} \tag{2.15}$$

where the parameter μ defines the center of the distribution and the parameter σ^2 defines both the height and the spread of the distribution. The smaller σ^2 is, the sharper the distribution becomes. In fact, μ and σ^2 correspond to the expected value and the variance of the distribution, respectively. These parameters are obtained by applying the first moment (mean) and the second moment (variance) to Eq. (2.15). The following expressions are obtained:

$$E[X] = \mu \quad \mathrm{VAR}[X] = \sigma^2 \tag{2.16}$$

Note from Fig. 2.5a that the highest probability density is always at the mean value of the normal distribution. Most of the data of the random variable tend to be

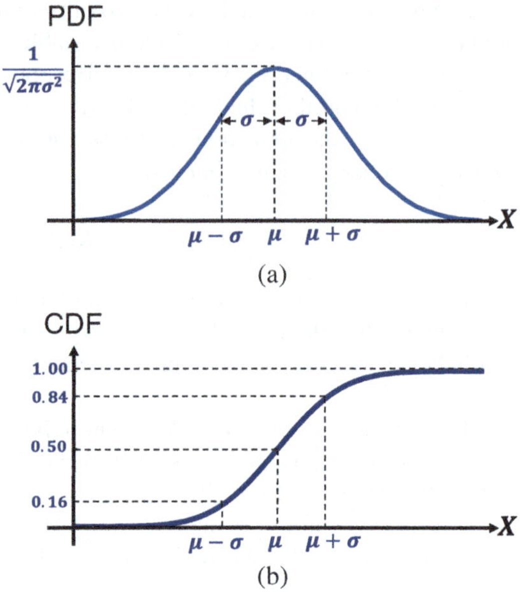

Fig. 2.5 (**a**) Probability density function (PDF) and (**b**) cumulative distribution function (CDF) of a normal-distributed continuous random variable

close to the mean, and gradually fewer data are found as the value of the random variable moves farther from the mean. Also, it is worth to notice that the normal distribution is symmetrical with respect to its mean value.

The shape of the cumulative density function (CDF) of a continuous random variable with a normal distribution is shown in Fig. 2.5b. Mathematically, it is given by [3, 8]

$$F(x) = \int_{-\infty}^{x} \frac{1}{\sigma\sqrt{2\pi}} \exp^{-\frac{(x-\mu)^2}{2\sigma^2}} dx \tag{2.17}$$

It does not exist an explicit expression for the CDF of a normal distribution (Eq. (2.17)). However, the CDF is usually computed using numerical integration methods. The CDF starts from zero and gradually increases to the unit as shown in Fig. 2.5b. Around 15% of the values that the random variable takes are below the point $\mu - \sigma$, a 50% of the values are below the mean, and 84% are below the $\mu + \sigma$ value. The rate of increase of the CDF near the mean value depends on the variance of the distribution (σ^2). When σ^2 is smaller, both the 15% and the 84% points are reached closer to the mean value.

Compared to the uniform distribution that has a well-defined interval, the normal distribution spread depends on the value of σ. Figure 2.6 shows probabilities associated with a normal distribution for bins with a width of 1σ. As can be observed, the probability of the random variable X to take a value far from the mean is very low. As can be observed, 68.2% of the outcomes will be located within the

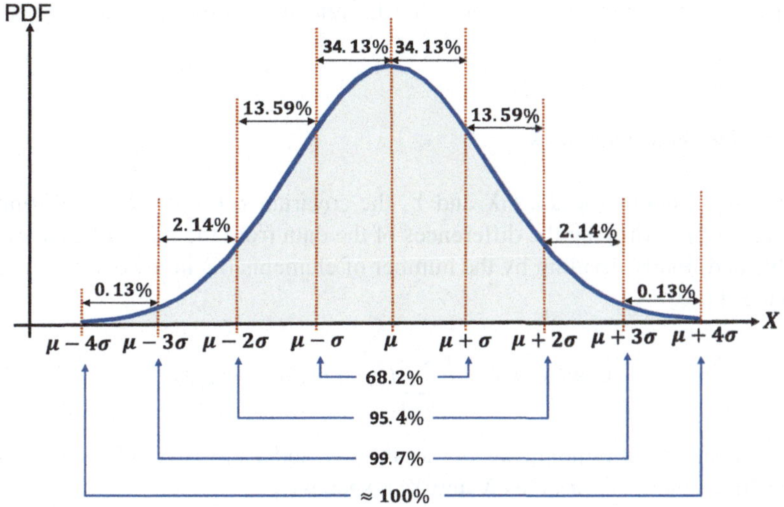

Fig. 2.6 Normal distribution with typical probabilities around the mean

$\mu \pm \sigma$ interval. A quick rule of thumb is to assume that the width of the distribution is of 6σ $(\mu \pm 3\sigma)$ since 99.7% of the possible values of the random variable are within this interval.

2.5 Relationship Properties of Random Variables

This section describes two important metrics, covariance and correlation, which allow quantifying the relationship between random variables.

2.5.1 Covariance

Covariance is a metric that allows knowing if two random variables are related between them [7–9]. In other words, covariance measures if the random variables are independent or dependent on each other. It measures the expected behavior of one variable with respect to the change of the other variable. The behavior of the data relationship between two random variables can be positive or negative. The relationship is positive when the value of both variables tends to increase or decrease in the same sense and is negative when the value of one variable tends to increase (decrease), while the value of the other variable tends to decrease (increase). Mutually independent variables do not keep a relationship between them; i.e., if the value (outcome) of one variable is known, it does not change the expected value of the other.

The computation of the covariance for discrete and continuous variables is given next.

2.5.1.1 Discrete Variables

For discrete random variables X and Y, the covariance is calculated by summing the results of multiplying the differences of the data from the mean of each random variable and finally dividing by the number of elements (n) in the dataset. It can be written as [7–9]

$$\mathrm{Cov}(X, Y) = \frac{1}{n}\sum_{i=1}^{n}(x_i - \mu_X)(y_i - \mu_Y) \tag{2.18}$$

where μ_X and μ_Y are the means of variables X and Y, respectively, and x_i and y_i are the ith element of variables X and Y, respectively.

From (2.18), it can be deduced that the covariance between the same variable is the variance of the variable; for instance, $\mathrm{Cov}(X, X) = \sigma_X^2$. The covariance is zero when the variables are mutually independent.

Example 3 Let X and Y be two discrete random variables with the following datasets, $X[2.1, 2.5, 3.6, 4.0]$ and $Y[8, 10, 12, 14]$. The covariance between X and Y is obtained using (2.18).

First, the mean of variables X and Y is calculated using (2.7). It gives

$$\mu_X = 3.05 \qquad \text{and} \qquad \mu_Y = 11$$

Then, the covariance is obtained as follows:

$$\mathrm{Cov}(X, Y) = \frac{1}{4}\sum_{i=1}^{4}(x_i - 3.05)(y_i - 11)$$

where x_i and y_i are values of variables X and Y, respectively. The result is

$$\mathrm{Cov}(X, Y) = 2.26$$

The resulting value means that both variables are positively related between them; i.e., if the value of one variable increases or decreases, the other one does it too.

2.5.1.2 Continuous Variables

For continuous random variables X and Y, the covariance is calculated by obtaining the difference between the expected value of the product XY ($E[XY]$) and the product of the means $E[X] = \mu_X$ and $E[Y] = \mu_Y$ of the respective variables. It can be written as [7–9]

$$\mathrm{Cov}(X, Y) = E[XY] - E[X]E[Y] \tag{2.19}$$

where the expected value $E[XY]$ is given by [7–9]

$$E[XY] = \int_a^b \int_a^b xyf(x, y)dxdy \tag{2.20}$$

$f(x, y)$ is a bivariate function that defines the relationship between the two continuous random variables. a and b are constants that establish the bounds of the bivariate function. $a < b$ and both can take values in the interval $(-\infty, \infty)$.

2.5.2 *Correlation*

Correlation is a metric that provides the linear relationship degree between two random variables. The correlation value is dimensionless and just takes values in the range of $-1 \leqslant \rho \leqslant 1$. For this reason, the correlation is considered the normalized version of the covariance. Therefore, if the resulting value of the covariance is positive or negative, the correlation value is also positive or negative, respectively, in the range of $-1 \leqslant \rho \leqslant 1$. Moreover, if the covariance is zero, the correlation is also zero. There are several ways to compute the correlation degree between two random variables; however, regarding the interest of this book, the correlation value is defined by Pearson's coefficient [1, 9].

The Pearson's coefficient is calculated by dividing the covariance of the concerning random variables (X and Y) with the product of the standard deviations (σ_X and σ_Y) of the same random variables, and it is expressed as [1, 9]

$$\rho_{XY} = \frac{\mathrm{Cov}(X, Y)}{\sigma_X \sigma_Y} \tag{2.21}$$

Figure 2.7 displays scatter plots with different correlation degrees between two random variables, X and Y. To each value of X corresponds a value of Y (one dot). Figure 2.7a exemplifies the case of two variables strongly correlated ($\rho = 1$). It is observed that the data relationship between the two variables is completely linear. Figure 2.7b shows the case of two variables with medium to high correlation ($\rho = 0.8$). It can be noted that the data relationship between both variables presents a semilinear behavior with dispersion. Figure 2.7c, d depict the behavior of two random variables with low ($\rho = 0.4$) and zero ($\rho = 0$) correlation between them. In both cases, $\rho = 0.4$ and $\rho = 0$, a disperse behavior of the data relation between variables is presented.

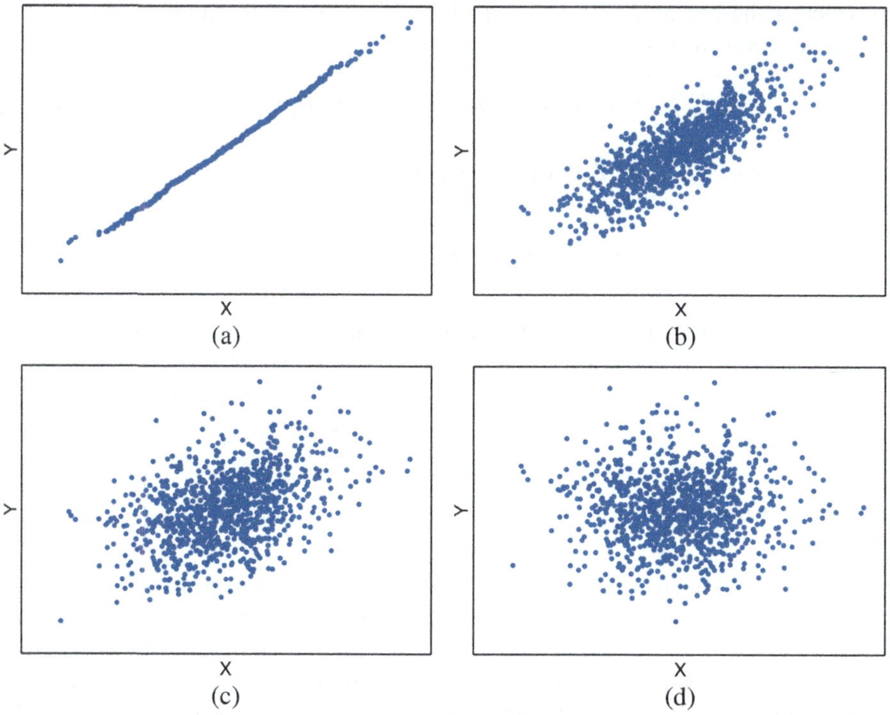

Fig. 2.7 Scatter plots with different correlation degrees between two random variables. (**a**) $\rho = 1$. (**b**) $\rho = 0.8$. (**c**) $\rho = 0.4$. (**d**) $\rho = 0$

For instance, the correlation between the two datasets of variables X and Y given in Example 3 is obtained using (2.21). The result is $\rho = 0.979$. The obtained correlation value indicates that both variables are strongly correlated between them, and both change toward the same sense.

2.6 Sum of Normal Random Variables

In this section, the sum of two or more normal random variables is described. The resulting sum is a random variable normally distributed with a mean and a variance.

2.6.1 Sum of Two Normal Random Variables

Let X_1 and X_2 be two random variables normal-distributed, each represented by $X_1 \sim N(\mu_{X_1}, \sigma^2_{X_1})$ and $X_2 \sim N(\mu_{X_2}, \sigma^2_{X_2})$. μ_{X_1} and μ_{X_2} are the mean of the variables X_1 and X_2, respectively. σ_{X_1} and σ_{X_2} are the variance of the variables X_1 and X_2, respectively. The sum of the two random variables can be expressed as

$$Y = X_1 + X_2 \tag{2.22}$$

The computation of the mean and variance of the resulting sum, represented by the variable Y, is described next [1, 3–5, 7, 9, 11].

2.6.1.1 Mean

The mean of Y is obtained by summing the mean of each random variable as follows [9, 11]:

$$\mu_Y = \mu_{X_1} + \mu_{X_2} \tag{2.23}$$

2.6.1.2 Variance

Variance calculation depends on whether the random variables X_1 and X_2 are related between them or not [1, 3–5, 7, 9, 11], i.e., if they are *independent* or not (see Sect. 2.5).

For *independent* normal random variables, the variance of Y is obtained by adding the variance of each random variable as follows [9, 11]:

$$\sigma_Y^2 = \sigma_{X_1}^2 + \sigma_{X_2}^2 \tag{2.24}$$

For *dependent* normal random variables, the variance of Y is given by [9, 11]

$$\sigma_Y^2 = \sigma_{X_1}^2 + \sigma_{X_2}^2 + 2\text{Cov}(X_1, X_2) \tag{2.25}$$

where $\text{Cov}(X_1, X_2)$ is the covariance between the two random variables, and it is calculated as described in Sect. 2.5.

2.6.2 *Sum of More than Two Normal Random Variables*

The sum of n normal random variables is merely an extension of the sum of two normal random variables. Let $X_1, X_2, \ldots, X_n$ be n random variables normal-distributed, each represented by $X_1 \sim N(\mu_{X_1}, \sigma_{X_1}^2)$, $X_2 \sim N(\mu_{X_2}, \sigma_{X_2}^2), \ldots,$ $X_n \sim N(\mu_{X_n}, \sigma_{X_n}^2)$.

Consider

$$Y = X_1 + X_2 + \cdots + X_n \tag{2.26}$$

The computation of the mean and variance of the resulting sum, represented by the variable Y, is described next [1, 3–5, 7, 9, 11].

2.6.2.1 Mean

The mean of Y for n random variables is obtained by summing the mean of each random variable as follows [9, 11]:

$$\mu_Y = \mu_{X_1} + \mu_{X_2} + \cdots + \mu_{X_n} \tag{2.27}$$

2.6.2.2 Variance

As mentioned previously, variance calculation depends on whether the random variables $X_1, X_2, \ldots, X_n$ are *independent* or *dependent* among them [1, 3–5, 7, 9, 11].

For *independent* normal random variables, the variance of Y for n random variables is obtained by adding the variance of each random variable as follows: [9, 11],

$$\sigma_Y^2 = \sigma_{X_1}^2 + \sigma_{X_2}^2 + \cdots + \sigma_{X_n}^2 \tag{2.28}$$

For *dependent* normal random variables, the variance of Y for n random variables is given by [9, 11]

$$\sigma_Y^2 = \sigma_{X_1}^2 + \sigma_{X_2}^2 + \cdots + \sigma_{X_n}^2 + 2\sum_{i=1}^{n-1} \sum_{j=i+1}^{n} \mathrm{Cov}(X_i, X_j) \tag{2.29}$$

where $\mathrm{Cov}(X_i, X_j)$ is the covariance between each pair of variables, and it is calculated as described in Sect. 2.5.

2.7 Series and Theorem of Taylor

The series and theorem of Taylor are commonly used to approximate functions by a sequence of terms. The series and theorem of Taylor can be applied to functions which depend on a single variable, two variables, and more than two variables. This section describes the basics to approximate a function (with a single and two variables) using the series of Taylor.

2.7.1 Basic Definitions

Some basic concepts are given next:

A series: It is the sum of a succession of terms which can represent a function [13, 14].

A power series: It is the sum of a succession of terms which can represent a function as a polynomial of degree n [13, 14].

A series of Taylor: It is the sum of a succession of terms which can represent a function infinitely differentiable about a given point [9, 12–15].

2.7.2 Single Variable

Let $f(x)$ be a function, with variable x, infinitely differentiable about a given point $x = a$. The function $f(x)$ can be approached by a sequence of terms using the *series of Taylor* as follows [9, 12–15]:

$$f(x) = f(a) + f'(a)(x-a) + \frac{f''(a)}{2!}(x-a)^2 + \cdots + \frac{f^n(a)}{n!}(x-a)^n \tag{2.30}$$

where $f(a)$ is the function $f(x)$ evaluated at a, $f'(a)$ represents the first derivative of the function evaluated at a, $f''(a)$ represents the second derivative of the function evaluated at a, and $f^n(a)$ represents the nth derivative of the function evaluated at a.

Example 4 Considering the function $f(x) = e^x$ about $x = a$, where $a = 0$. Obtain the expanded expression in series of Taylor for $n = 3$.

It is known that for $f(x) = e^x$, the n derivatives of the function results the same function, and their respective evaluations at zero are 1s. Hence, using (2.30), the Taylor's series (polynomial of third order) that approximates the function $f(x) = e^x$ results in the following:

$$e^x = 1 + x + \frac{x^2}{2!} + \frac{x^3}{3!}$$

According to the obtained series, the function $f(x) = e^x$ around zero can be expressed as

$$e^x = \sum_{n=0}^{\infty} \frac{x^n}{n!}$$

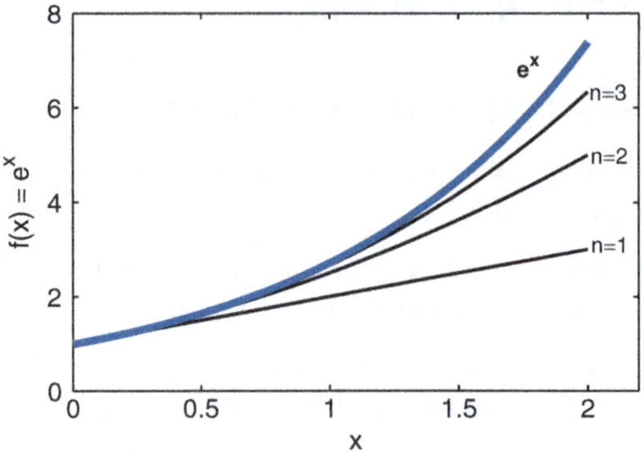

Fig. 2.8 Taylor series approximation of the function $f(x) = e^x$

Figure 2.8 plots the function $f(x) = e^x$ and the first three terms of the series of Taylor. It shows that Taylor's series utterly approach the function $f(x) = e^x$ with few terms. It is observed that the larger the value of n is, the higher the accuracy of the approximation.

2.7.3 *Two Variables*

Let $f(x_1, x_2)$ be a function, with variables x_1 and x_2, infinitely differentiable about the points $(x_1 = a_1, x_2 = a_2)$. The function $f(x_1, x_2)$ can be approached by a sequence of terms using the *series of Taylor* as follows [9, 12–15]:

$$\begin{aligned} f(x_1, x_2) = {} & f(a_1, a_2) + f'_{x_1}(a_1, a_2)(x_1 - a_1) + f'_{x_2}(a_1, a_2)(x_2 - a_2) \\ & + \frac{1}{2!}[f''_{x_1x_1}(a_1, a_2)(x_1 - a_1)^2 + f''_{x_1x_2}(a_1, a_2)(x_1 - a_1)(x_2 - a_2) \\ & + f''_{x_2x_2}(a_1, a_2)(x_2 - a_2)^2] \end{aligned} \tag{2.31}$$

where $f(a_1, a_2)$ is the function $f(x_1, x_2)$ evaluated at $x_1 = a_1, x_2 = a_2$, f'_{x_1} denotes the first derivative of the function with respect to the variable x_1 evaluated at $(x_1 = a_1, x_2 = a_2)$, f'_{x_2} represents the first derivative of the function with respect to the variable x_2 evaluated at $(x_1 = a_1, x_2 = a_2)$, and so forth.

More information about the approximation of functions (with two and more than two variables) with the series of Taylor can be found in Appendix A.

2.8 Summary

In this chapter, basic concepts of the probabilistic and statistical theory were given. First, discrete and continuous random variables are described along with its distribution functions, followed by an explanation of their characteristics, mean, and variance. The uniform and normal distributions that model the behavior of a random variable were presented. The relationship properties of random variables such as the covariance and correlation are discussed. Also, the sum operation for two and more than two random variables, independent and correlated, was presented. Finally, this chapter gives the fundamentals of the series and the theorem of Taylor (single and two variables) which are the basis for the mathematical formulation of the statistical delay of a gate.

References

1. D.C. Montgomery, G.C. Runger, *Applied Statistics and Probability for Engineers*, 3rd edn. (Wiley, New York, 2003)
2. M. Lefebvre, *Basic Probability Theory with Applications* (Springer, New York, 2009)
3. C.E. Weatherburn, *A First Course in Mathematical Statistics* (ELBS/Cambridge University Press, London/Cambridge 1961)
4. F. Dekking, C. Kraaikamp, H. Lopuhaa, L. Meester, *A Modern Introduction to Probability and Statistics - Understanding Why and How* (Springer, London, 2005)
5. P.L. Meyer, *Introductory Probability and Statistical Applications* (Addison Wesley, Reading, 1970)
6. J.J. Kinney, *A Probability and Statistics Companion* (Wiley, New York, 2009)
7. S. Ross, *A First Course in Probability*, 8th edn. (Pearson, Boston, 2010)
8. P.G. Hoel, *Introduction to Mathematical Statistics* (Wiley, New York, 1966)
9. A. Papoulis, S.U. Pillai, *Probability, Random Variables and Stochastic Processes*, 4th edn. (McGraw-Hill, New York, 2002)
10. M.G. Bummer, *Principles of Statistics* (Dover Publications, New York, 2003)
11. T.T. Soong, *Fundamentals of Probability and Statistics for Engineers* (Wiley, New York, 2004)
12. S.M. Kay, *Intuitive Probability and Random Processes Using MATLAB* (Springer, New York, 2006)
13. A. Jeffrey, H.-H. Dai, *Handbook of Mathematical Formulas and Integrals*, 4th edn. (Elsevier, Amsterdam, 2008)
14. R. Wrede, M.R. Spiegel, *Advanced Calculus*. McGraw-Hill Schaum Series, 3rd edn. (McGraw-Hill, New York, 2010)
15. F. Hussain, Understanding Calculus - Online book. http://www.understandingcalculus.com/, October 2017

Chapter 3
Process Variations

Abstract The variation of the technological parameters through the different transistors and interconnections built during the manufacturing process has become an important issue with impact on the performance of nanometer digital circuits. As semiconductor technologies scale down their feature sizes, process variation effects aggravate resulting in circuits with a lower yield, overdesigned and susceptible to reliability issues. Designers must be acquainted with the sources of variation and their behavior to develop techniques that fulfill specifications efficiently. In this chapter, a brief overview of the main steps to fabricate circuits in CMOS technology is given. Afterward, the most important sources of process variations due to non-idealities of the manufacturing process are identified and described. Then, process variations are classified as systematic and nonsystematic variations. The latter is further divided into two types of variations, inter-die and intra-die variations. The impact of the different types of variations on transistor behavior is analyzed. Then, correlation models to quantify the spatial correlation due to a certain type of process variations are presented. Finally, modeling of the process variations is depicted.

3.1 Introduction

The feature size of semiconductor technologies has scaled down in order to increase the integration density per area unit and to enhance the performance of the chip. Process variations on device parameters have always been present, but their impact was limited in older technologies. It was not until the introduction of the 0.35 μm technology node that process variations became a serious problem in the performance of the circuit. As a result of the continuous shrinking of feature sizes, design-related parameter variations have increased making the task of the designer difficult. Nowadays, process variations are a major concern in the design of integrated circuits (IC), and this concern increases as the size of devices takes values of a few nanometers.

V. Champac, J. G. Gervacio, *Timing Performance of Nanometer Digital Circuits Under Process Variations*, Frontiers in Electronic Testing 39,
https://doi.org/10.1007/978-3-319-75465-9_3

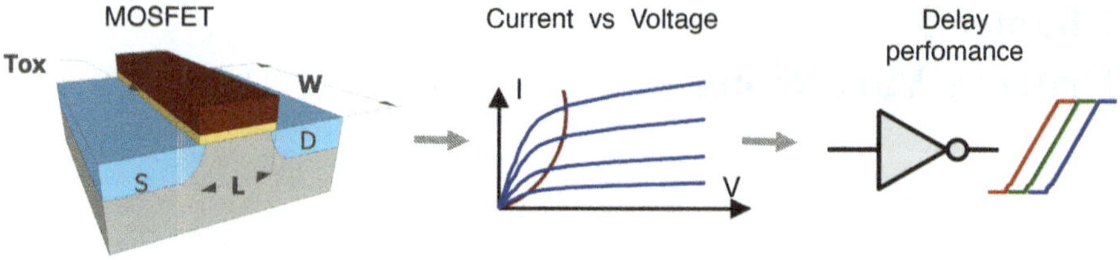

Fig. 3.1 The causal relation of the impact of process parameter variation. Physical parameter variations result in electrical parameter variations and the latter impact the circuit performance variations

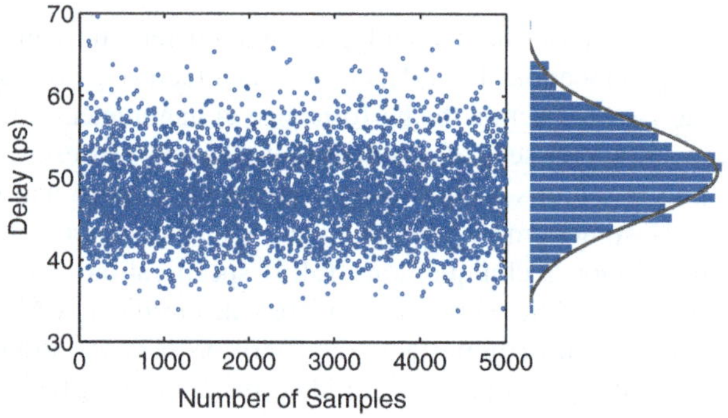

Fig. 3.2 Delay variability of a single logic gate due to process variations

Definition Process variations are the physical deviation of the parameters of devices and interconnections with respect to the intentionally designed values.

Fluctuations in the manufacturing process result in variations in the physical parameters that define the behavior of a circuit. Figure 3.1 shows a scheme of the cause and effect of parameter variations in integrated circuits. Variations in the physical parameters of devices and interconnections (channel length, channel width, gate oxide thickness, wire width and thickness) result in variations in their electrical characteristics (drive current, threshold voltage, wire resistance and capacitance). They produce performance variations at gate level (gate delay, slew rate, wire delay, power consumption), which translates into variations of performance at circuit and system level.

Figure 3.2 exemplifies the impact of process variations of physical parameters of devices on the performance of a single logic gate. The gate delay has a wide dispersion, and its value oscillates between 35 and 70 ps for all the considered samples. Note that frequency of occurrence of delay values approach to a normal Gaussian distribution.

3.2 CMOS Manufacturing Process

3.2.1 CMOS Technology Overview

The CMOS (complementary metal-oxide semiconductor) technology is the semiconductor technology widely used to build integrated circuits, such as microprocessors, memories, and application-specific integrated circuits (ASICs) [1–3]. This technology is the basis of modern electronics and has been capable of following the Moore's law for five decades [4]. An integrated circuit consists of a few to billions of transistors manufactured and interconnected on a semiconductor substrate, also known as a wafer. In order to incorporate such devices into an integrated circuit, specifications are translated into a description of layers, which are known as circuit layout. The layout design is the process of creating a physical representation of the devices and interconnections, and it complies with constraints imposed by the manufacturing technology [5]. The generation of such layout is usually accomplished by using computer-aided design (CAD) tools. In general, the circuit layout is subjected to functional and electrical verification before it is released for fabrication.

The complete manufacturing flow of an IC can be classified depending on the stage of the process as front end of line (FEOL) and back end of line (BEOL) [6]. FEOL refers to the steps toward the development of devices. BEOL relates to the steps to build the interconnections as well as any passive device for the circuit. Figure 3.3 shows a cross-section of a CMOS circuit where the FEOL and BEOL sections are shown.

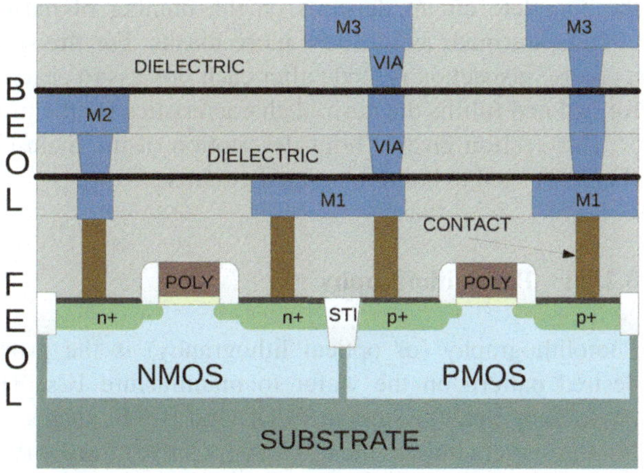

Fig. 3.3 Cross-section of a CMOS circuit illustrating the FEOL and BEOL sections

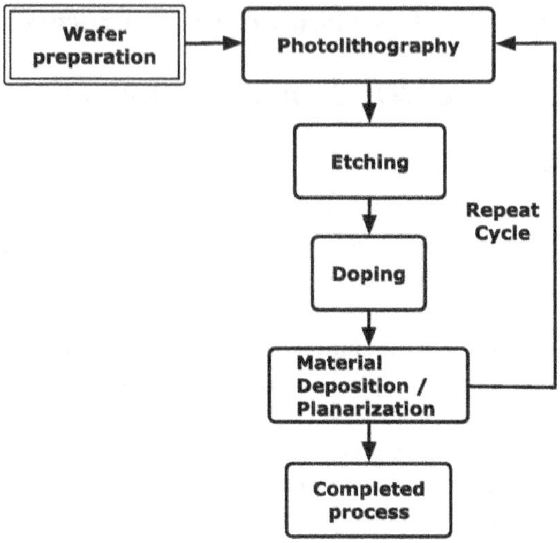

Fig. 3.4 Simplified manufacturing process flow

3.2.2 *Main Manufacturing Processes*

Throughout the manufacturing of an integrated circuit, a large number of processes are repeatedly performed until it is completely built, first the FEOL stage and then the BEOL [7, 8]. A simplified flow of the manufacturing process steps is shown in Fig. 3.4. The flow starts with the wafer preparation followed by the photolithography, etching, doping, material deposition, and planarization. These processes are repeated until the whole circuit is developed. The complexity to create the complete circuit depends on the number of masks used. Modern integrated circuits are made with 30 or more masks. For the success of the manufacturing, a precise inspection is made after each process to ensure the structure was correctly formed and fulfills the desired characteristics for the proper device operation [3, 7].

This section gives a brief description of the main manufacturing process steps (Fig. 3.4) used to build integrated circuits.

3.2.2.1 Photolithography

Photolithography (or optical lithography) is the process that allows creating a desired pattern on the wafer to manufacture ICs. The main steps used in the photolithography process at FEOL and BEOL stages are the following [7–10]: (a) Photoresist coating application. A (positive) film is deposited over the wafer, and in this film, the patterns are transferred from the mask. (b) Mask alignment. The mask is aligned with the wafer to transfer the patterns onto it. (c) Exposure of the wafer. The wafer covered with photoresist is exposed to ultraviolet (UV) light in those parts that are desired to remove the material. The mask defines the pattern. (d) Development. In this step, areas of the photoresist film exposed to the UV light

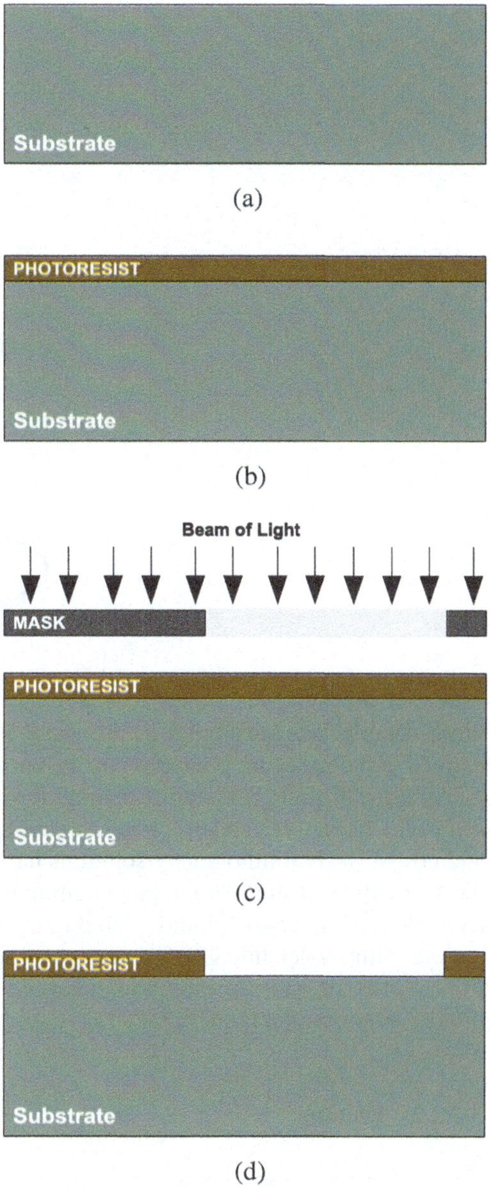

Fig. 3.5 Photolithography process steps. (**a**) Substrate. (**b**) Photoresist coating. (**c**) Exposure to UV light. (**d**) Development

are dissolved with a solution. (e) Hard baking. Finally, the wafer is baked to affirm the photoresist film and condition the wafer for subsequent processes. Figure 3.5 illustrates those photolithography process steps.

Figure 3.6 illustrates a basic schematic of a photolithography system which is constituted by an illumination system, an optical system, and the photomask; the latter carries the patterns to be drawn on the wafer. The illumination system

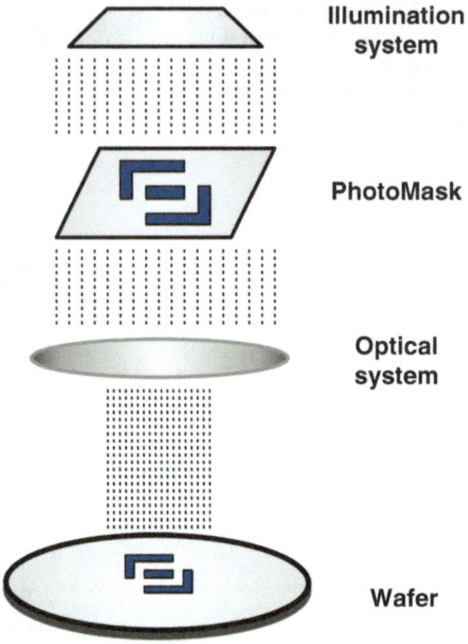

Fig. 3.6 Basic schematic of a photolithography system

generates the light that shines over the mask that contains the designed patterns; those patterns are transferred to the wafer by means the optical system. Patterns defined in the mask are printed on the layers of the wafer.

As a result of the continuous scaling down of the feature sizes of devices, at the scale of nanometers, modern photolithography systems have become very complex and costly. Several lithography solutions have successfully emerged to accomplish the fabrication of modern integrated circuits. Some of these are double patterning technology, immersion liquid with higher refraction index, spacer lithography, extreme ultraviolet lithography, nano-imprint lithography, and X-ray lithography [3, 7, 8, 11], as well as resolution enhancement techniques (RET) such as optical proximity correction (OPC) [3, 7, 11].

3.2.2.2 Etching

The etching process consists of removing the portion of the material that is deposited or grown on the wafer but was not defined in the mask [3, 7, 8]. Because barrier layer (oxidation) formation or material deposition extends over the whole wafer, the etching process is required to remove the non-desired part. Once the photolithography process draws the desired pattern on the wafer, the etching removes the remaining portion. Usually, the etching process is performed using chemical solutions that remove the unprotected parts while the photoresist film defined by the photolithography process acts as a protection. There are several

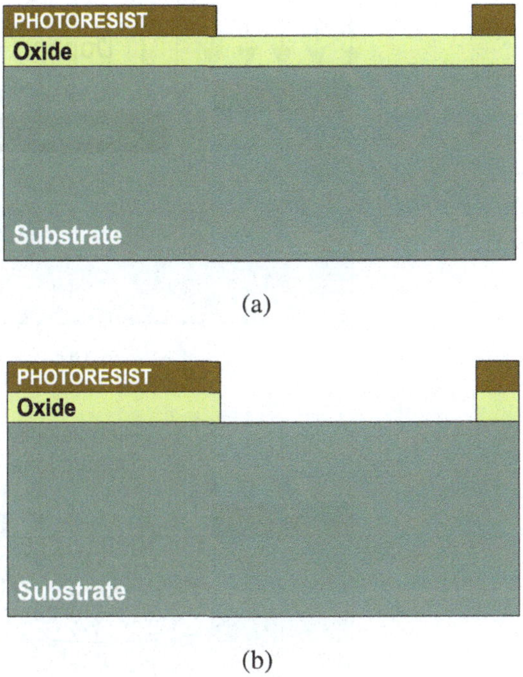

Fig. 3.7 Etching process. (**a**) Definition of the desired pattern with the mask. (**b**) Etching of oxide

etching techniques to perform this task; these are wet and dry etching, or plasma etching, and plasma sputtering [3, 7]. Figure 3.7a shows a cross-sectional view of a substrate with the deposited material (oxide) and the photoresist which define the desired pattern. Figure 3.7b illustrates the etched wafer where the non-desired material was removed. After the etching process, the photoresist is also removed.

3.2.2.3 Doping

Doping is the process of adding atoms into a material to modify its electrical (conduction) properties. *Diffusion* and *ion implantation* are the two most common methods in semiconductors to doping a material [3, 7, 8, 12]. With either diffusion or ion implantation, it is possible to place impurities or dopants (such as phosphorus, arsenic, boron, etc.) into the silicon wafer. The threshold voltage adjustment, source and drain formation, and well implantation, among others, are achieved by these methods. Figure 3.8 illustrates both doping techniques. Below, each one is described.

Diffusion is applied in two steps, firstly *pre-deposition* and then *drive-in diffusion*. In the pre-deposition step, dopants are deposited superficially into the substrate, and then dopants are carried deeper into the substrate with drive-in diffusion [3, 7, 8, 12].

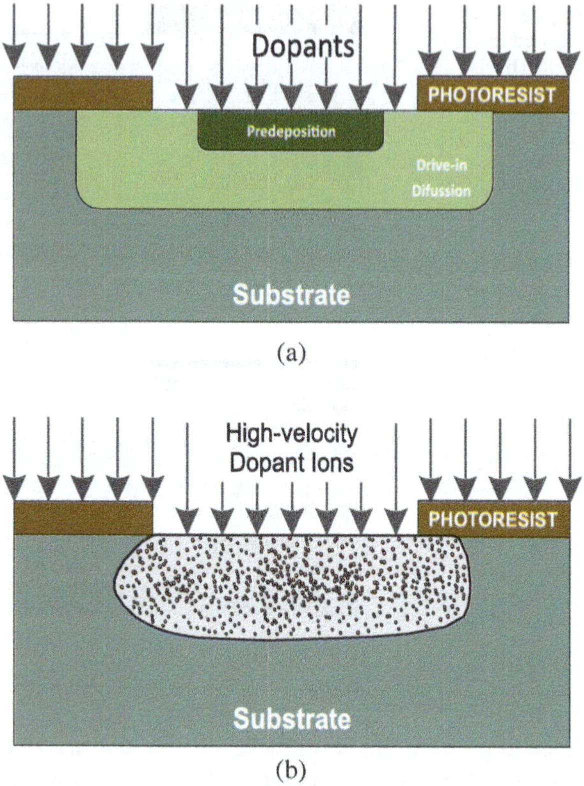

Fig. 3.8 Doping methods to introduce impurities or dopants into the silicon. (**a**) Diffusion. (**b**) Ion implantation

Ion implantation is performed by means of a flurry of ions accelerated at high velocities into the substrate [3, 7, 8, 12]. Ions implanted require being activated which is performed through an annealing process. The way to develop this doping process causes the number and placement of atoms implanted be completely random.

3.2.2.4 Deposition

The deposition process is the means by which a material is deposited over the wafer to create a thin layer. The deposition of material is always laid over the whole wafer. Some commonly deposited materials are dielectrics, polysilicon, and metals. The material can be deposited using either chemical or physical methods, known as chemical vapor deposition and physical vapor deposition [3, 7, 8, 13]. Other deposition methods are atmospheric-pressure chemical vapor deposition and atomic layer deposition [3, 7], which are capable of creating thin layers of diverse materials

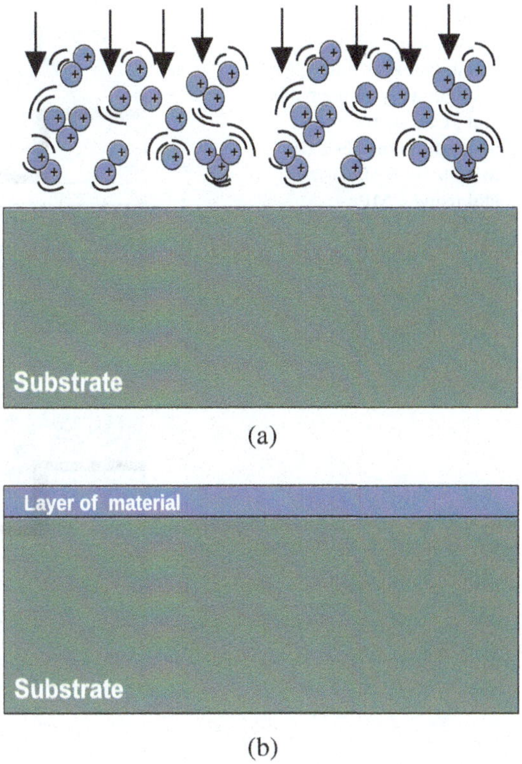

Fig. 3.9 Deposition process of a material over the substrate. (**a**) Deposition of a material over the substrate. (**b**) Thin layer created over the substrate

in modern semiconductor processes. Figure 3.9a illustrates the deposition process of particles of material over the substrate to create the desired layer. Figure 3.9b shows the layer created over the substrate.

3.2.2.5 Planarization

The process of planarization is the action of obtaining a flat and smooth surface out of a rough topography. Planarization is necessary when manufacturing integrated circuits because the deposition of different materials (during the formation of the front-end- and back-end-of-line stages) results in nonuniform topographies [3, 7, 8]. This action is necessary after each layer of material is deposited, so that formation of next layer begins from a surface as flat as possible.

There are several planarization techniques; these are thermal flow, photoresist etchback, sputtering etchback, spin-on-glass etchback, and chemical mechanical polishing (CMP) [3, 7, 8]. The latter is the planarization technique currently used because it allows getting a higher resolution of the photolithography process, thus enabling the continuing scaling down [14–16]. The CMP technology was introduced with the 0.25 μm technology node and is mostly used during the creation of dielectrics and interconnections. Figure 3.10 exemplifies the CMP-

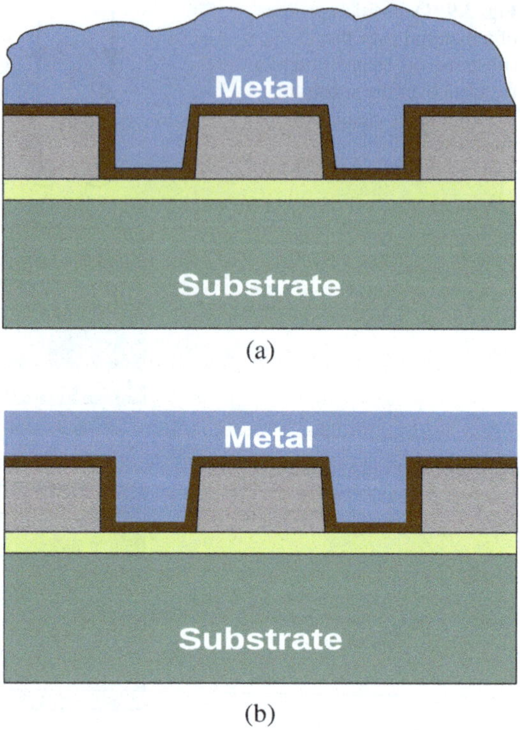

Fig. 3.10 Planarization process using chemical mechanical polishing (CMP). (**a**) Substrate with deposited metal previous the planarization process. (**b**) Substrate with planarized metal using CMP

based planarization process. Figure 3.10a shows a substrate with deposited metal previous to the planarization process. After CMP is applied, the surface takes a flat topography (Fig. 3.10b).

3.3 Sources of Process Variations

3.3.1 Sources of Variation on Device Parameters

Physical and electrical parameters of devices define the performance characteristics. The channel length (L), the channel width (W), oxide thickness (T_{ox}), and threshold voltage (V_{th}) are the main parameters that define the performance behavior of MOSFET devices. Therefore, variations of the nominal values of these parameters due to the manufacturing process change the expected performance profits [6, 17–19]. The channel length (L) and width (W) are physical parameters that determine the dimensions of a MOSFET transistor and are known as the *critical dimensions* (CD). The dielectric that isolates the gate electrode from the substrate has a defined thickness (T_{ox}); this physical parameter plays a critical role in the behavior of the device. The threshold voltage (V_{th}) is the minimum voltage required at the transistor gate to turn it on. The principal sources of variation of these parameters are outlined below.

3.3.1.1 Sources of Variation in the Channel Length

The channel length (L) is the smaller critical dimension of the transistor; it defines the size of the technology. Effective channel length depends on a large number of process steps and equipment used during manufacturing [3]. However, the photolithography and etching process steps produce the most important channel length variations across the chip [6, 18, 20–22]. Line edge roughness [23] and optical proximity effect [24] are two important sources of variations strongly impacting the effective channel length [24]; both are described below.

3.3.1.2 Line Edge Roughness

Line edge roughness (LER) is the random variation of the polysilicon edges along its width. Because the photolithography systems use light sources with wavelengths larger than the minimum feature size (critical dimension), lithography-related process leads to larger gate variation due to LER [20, 25, 26]. LER has become an important source of variation of channel length, and it is expected to continue rising in coming technologies [27]. Figure 3.11 illustrates the LER on a polysilicon gate of a transistor. It can be observed that the effective channel length fluctuates along the channel width.

3.3.1.3 Optical Proximity Effect

Optical proximity effect (OPE) is the variation in the size and shape of a feature as a function of the proximity of other neighboring features [22, 24, 26]. OPE occurs during the photolithography process due to the diffraction of light with neighboring structures. Because of this, the size and shape of an isolated line would be different from the size and shape of a line in an array. In order to

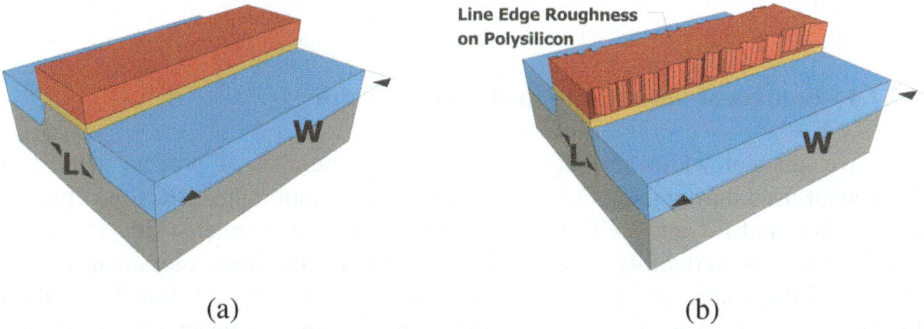

Fig. 3.11 Line edge roughness (LER) exemplification on a polysilicon gate of a transistor. (**a**) Polysilicon gate w/o LER. (**b**) Polysilicon gate with LER

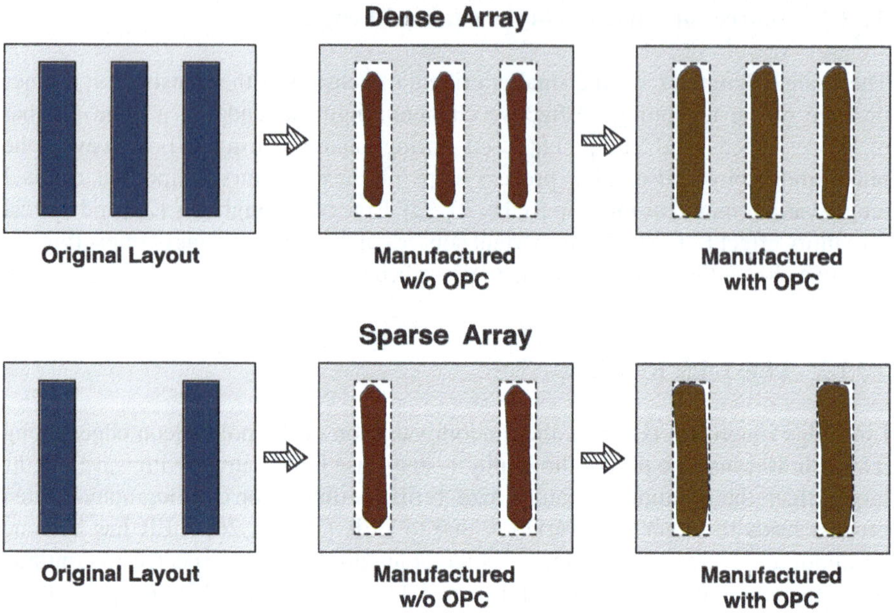

Fig. 3.12 Impact of optical proximity effect on a sparse and dense array of lines and its attenuation with the optical proximity correction technique

overcome this photolithography issue, a resolution enhancement technique called optical proximity correction (OPC) is used [24]. Figure 3.12 typifies two cases of an array of lines, sparse and dense, both with the impact of OPE and using OPC. In a sparse array, OPE notably changes the size and shape of manufactured lines; however, manufactured lines closely approach the original layout when the OPC technique is used. In a dense array, OPE changes the size and shape of manufactured lines significantly; however, manufactured lines approach well to the original layout when the OPC technique is used. OPE has a greater impact on a dense array than a sparse array.

3.3.1.4 Sources of Variation in the Channel Width

Effective channel width also depends on a large number of process steps and equipment used during manufacturing. Sources of variation that affect the channel length such as LER and OPE also impact the effective channel width. However, other sources of variability of the channel width are the mask alignment during the photolithography process and the formation of the shallow trench isolation [6, 19, 28]. The channel width variability has a lower impact on the device performance than channel length variability because the minimum channel width size is larger than the minimum channel length size.

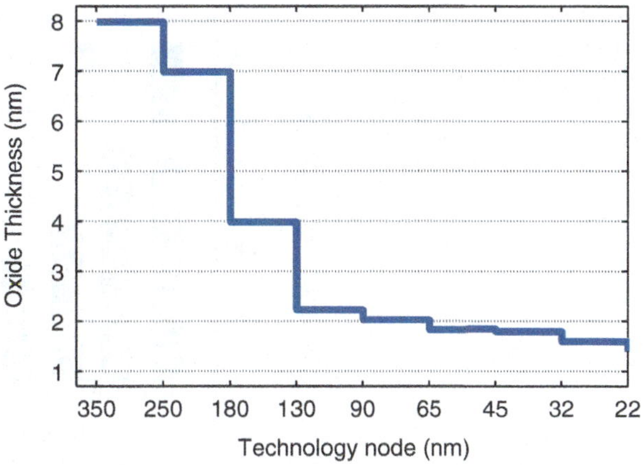

Fig. 3.13 Trend of the gate oxide thickness (T_{ox}) as the feature size of the technology scales down [29]

3.3.1.5 Sources of Variation in the Gate Oxide Thickness

Silicon dioxide (SiO_2) is the dielectric material that separates the gate electrode from the channel region of an MOS transistor [3, 8]. As feature sizes of devices are scaled down to obtain chips with better performance, the oxide thickness (T_{ox}) of the transistor is also reduced (Fig. 3.13). In earlier technologies, T_{ox} was in the order of tens of nanometers, but in modern technologies, it has just a few nanometers or angstroms [6, 29]. Although the dielectric layer can be grown with high accuracy, a small variation can have a significant impact on the device's electrical properties. Besides, a thinner dielectric introduces quantum effects. For this reason, new materials with higher dielectric constant have appeared [30, 31].

3.3.1.6 Sources of Variation in the Threshold Voltage

The threshold voltage (V_{th}) value of a MOSFET transistor is determined by several characteristics during its manufacturing; these include the gate electrode material, the thickness of the dielectric, the carriers' concentration, and the doping profile in the channel, among others [6, 17–19, 32]. Therefore, any alteration in each of these characteristics results in a change of the desired threshold voltage of the transistor. The primary source of V_{th} variation is due to the random dopant fluctuation during the doping process; this variability source is described next.

3.3.1.7 Random Dopant Fluctuation

The channel region of the transistor is doped with atoms of a material different to the substrate to control and determine the threshold voltage. This process is achieved via ion implantation (see Sect. 3.2.2). The way to develop this process causes that the amount and location of atoms implanted in the channel of the transistor to be completely random; thus the term *Random Dopant Fluctuation* (RDF) [3, 33].

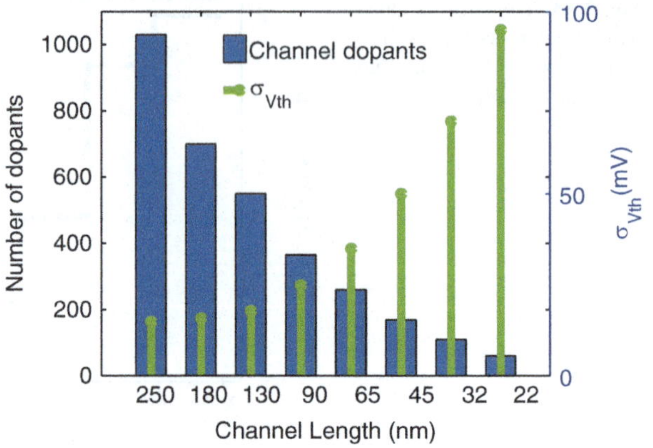

Fig. 3.14 Behavior of threshold voltage deviation as shrinking feature size while the number of dopants placed into the channel decreases [34]

Because of the random nature of the carriers' concentration, the V_{th} shift is different and completely random from transistor to transistor. Figure 3.14 illustrates that the threshold voltage deviation ($\sigma_{V_{th}}$) increases when shrinking feature size while the number of dopants placed into the channel decreases [6, 34]. This tendency imposes a serious challenge for the circuit designers and for the circuit performance itself.

The standard deviation of threshold voltage variation for a transistor with minimum dimensions is defined by Pelgrom's model [6, 28, 35] as follows:

$$\sigma_{V_{th}0} = \frac{qT_{ox}}{\varepsilon_{ox}}\sqrt{\frac{N_a W_d}{3L_{min}W_{min}}} \tag{3.1}$$

where T_{ox} is the oxide thickness, ε_{ox} is the oxide permittivity, N_a is the channel doping, W_d is the depletion region width, and L_{min} and W_{min} are the minimum channel length and width, respectively.

According to Poisson statistics [36, 37], fluctuations in the number of dopant atoms have a standard deviation equal to the square root of the number of dopants. Therefore, the threshold voltage deviation accounting for the effect of RDF is simply a function of its gate dimensions and its threshold voltage deviation ($\sigma_{V_{th}0}$) at minimum dimensions of the technology (defined by (3.1))[6, 17–19]. It is given by

$$\sigma_{V_{th}} = \sigma_{V_{th}0}\sqrt{\frac{L_{min}W_{min}}{L_{eff}W_{eff}}} \tag{3.2}$$

where L_{eff} and W_{eff} are the effective gate dimensions of the transistor.

3.3.2 *Sources of Variation in Interconnections*

Many of the processes used for device manufacturing are also used to build interconnections. Hence, several of the sources of process variations that affect the development of devices also impact interconnections, particularly those related to lithography and etching. Nevertheless, another important back-end-of-line process that is a source of process variations is the CMP-based planarization. The main CMP-related issues are described below.

3.3.2.1 Chemical Mechanical Polishing

Chemical mechanical polishing (CMP) is the planarization technique currently used during the back-end-of-line manufacturing process [14–16, 38]. The CMP process has the purpose of leaving the surface as flat as possible before the next layer is deposited. Unfortunately, CMP introduces two adverse effects that alter the shape of interconnections and consequently change the performance of the circuit; these are *dishing* and *erosion* [6, 15]. A schematic representation of these anomalies is illustrated in Fig. 3.15.

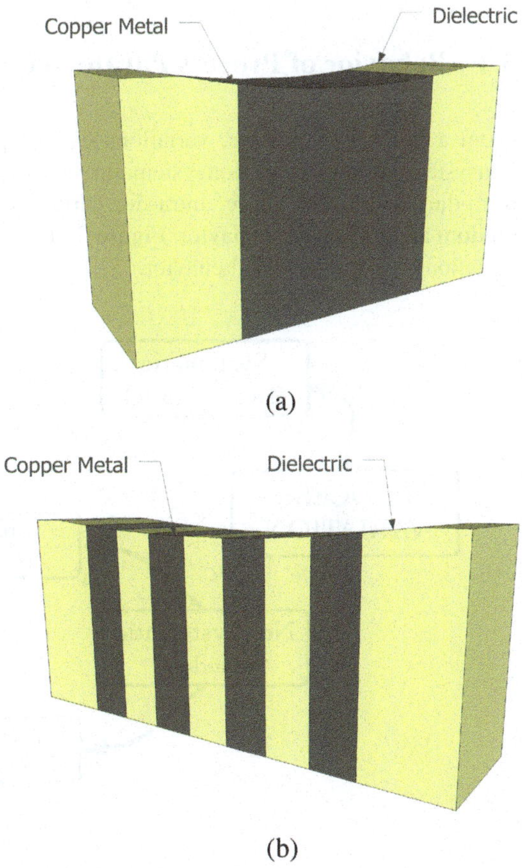

Fig. 3.15 Planarization issues due to CMP. (**a**) Dishing. (**b**) Erosion

Such effects are defined as follows:

Dishing is a cavity within an interconnect; it is the loss of metal as a result of overpolishing [6, 15]. This issue occurs in wide interconnect metals that are surrounded by a dielectric. This fact is because copper metal is a softer material than dielectric; therefore wide interconnections suffer overpolishing.

Erosion is the loss of both interconnect and dielectric as a result of overpolishing [6, 15]. This issue occurs in thin interconnect lines and areas with dense metallization. This fact is because copper metal is a softer material than dielectric; therefore, areas with dense metal lines are polished faster than areas with sparse metal lines.

Both interconnect issues result in an increment of metal resistance. Also, both can affect the development of upper levels of interconnection and consequently can introduce defects and malformed interconnections. In general, dishing and erosion can degrade the reliability of the chip.

3.4 Behavior of Process Parameter Variations

From a global perspective, variations can be produced due to a systematic and nonsystematic behavior. Nonsystematic variation can be generated at inter-die and intra-die scale. Even more, intra-die variation can be produced due to a purely random and correlated behavior. Figure 3.16 illustrates the classification of process variations according to its behavior.

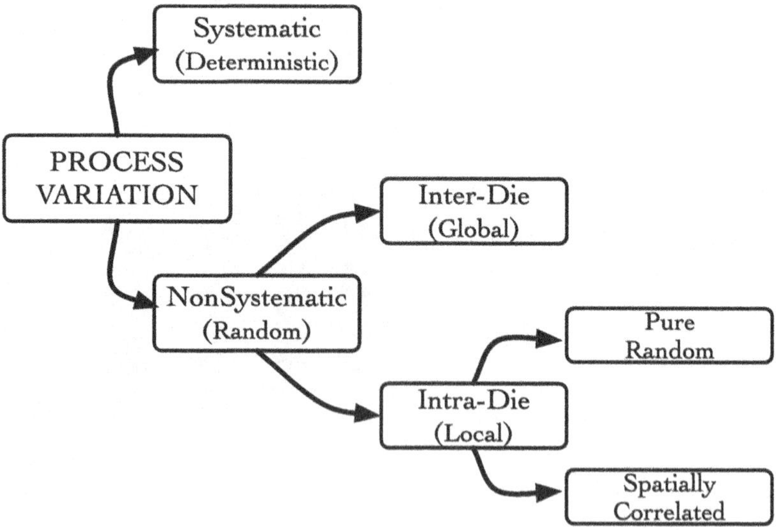

Fig. 3.16 Classification of process variations according to its behavior. Adapted from [39]

3.4.1 Systematic

Systematic (deterministic) variations are those that are produced systematically by the same steps of the manufacturing flow. In the development of devices and interconnections, several processes can cause the systematic variations; however, the most important is the photolithography process [18, 40]. Optical proximity effects (see Sect. 3.3) in nearby patterns are the primary source of these variations; therefore, most of this kind of variation is related to the layout design. Due to the nature of systematic variation, the amount of introduced variation is approximately the same. Because of this, it can be predictable and correctable using resolution enhancement techniques such as OPC. Some remaining may exist after OPC [20, 24, 26].

3.4.2 Nonsystematic

Nonsystematic (random) variations are those that are produced randomly from a large number of process steps and equipment used in the manufacturing flow [19, 39, 41]. In the development of devices and interconnections, many processes can cause random variations. Due to their arbitrary nature, random variations are statistically modeled and are the most important from a designer point of view. This kind of variations can be divided according to their behavior as inter-die (global) and intra-die (local) variations. Both kinds of variations are described below.

3.4.2.1 Inter-die Variations

Inter-die (D2D) variations (or global or die-to-die variations) affect all manufactured devices of a chip in the same way [39, 41–43]. The value of a parameter in all devices in a chip has the same value due to D2D variations. Inter-die variations occur at different levels of the manufacturing process as Fig. 3.17 illustrates. (a) Die-to-die variations can occur during the exposure of one mask to the next because the scanner has to be aligned to the previously realized process [19, 39, 43]. (b) Wafer-

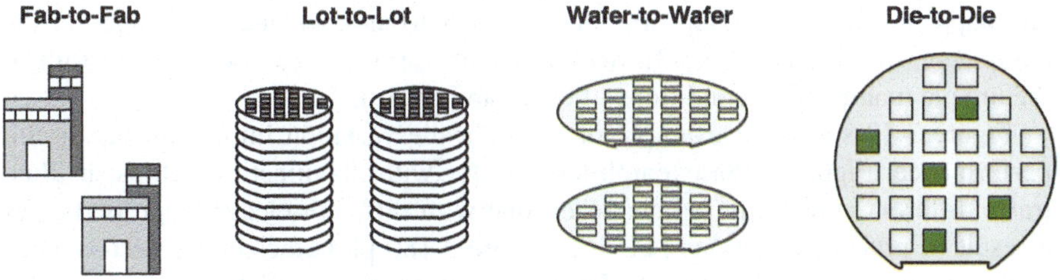

Fig. 3.17 Inter-die variation components

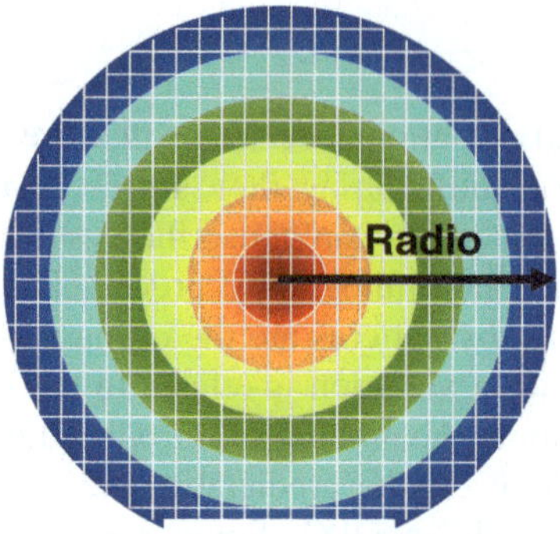

(a)

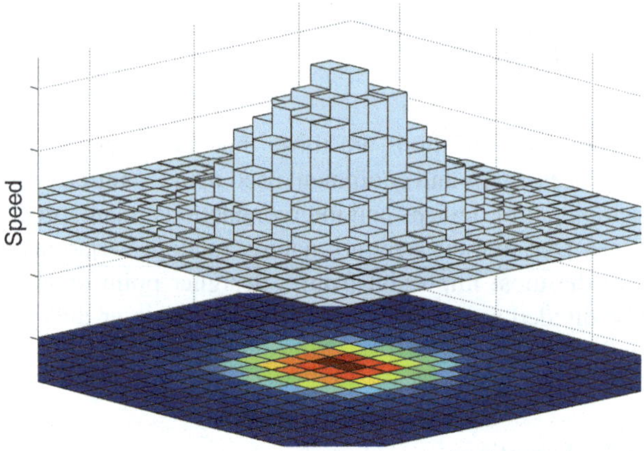

(b)

Fig. 3.18 Inter-die variation. In the bottom plot, the speed magnitude of the chips is mapped in concentric regions. In the top plot, the bars represent the speed magnitude of the chips in the corresponding location on the wafer. (**a**) Wafer with manufactured chips. (**b**) The speed magnitude of the manufactured chips

to-wafer variations can take place when processing one wafer to another due to the inexact position of the wafers in the equipment [19, 39, 43]. (c) Lot-to-lot variations can happen when processing one lot of wafers to another due to changes in the equipment working [19, 39, 43]. (d) Fab-to-fab variations can occur as a result of the unique manufacture conditions of each fab [19, 39].

Figure 3.18 shows an example of the inter-die variation impact on the circuit performance. Figure 3.18a exemplifies a wafer with manufactured chips; it illustrates different concentric regions, each one with different electrical properties as a result of process variations [44–46]. Figure 3.18b plots the timing performance of the wafer chips. In the bottom plot, the speed magnitude of the chips is mapped in concentric regions. In the top plot, the bars represent the speed magnitude of the chips in the corresponding location on the wafer. Those chips located at inner concentric regions of the wafer are in average faster than the ones placed at the edge [44–46]. This performance behavior indicates that physical parameter variations of devices in a chip are strongly correlated due to D2D variation.

3.4.2.2 Intra-Die Variations

Intra-die (WID) variations (or local or within-die variations) affect each manufactured device in the same chip differently. WID variations cause the feature sizes of some devices in a chip to have smaller dimensions than nominal, whereas other devices of the same chip have larger dimensions than nominal [39, 41–43]. Because of this, devices designed with equal size and shape, and manufactured in the same die, result in intrinsic properties different from each other and consequently with different performance. At subnanometer technologies, intra-die variation represents the largest part of the total variation. Due to the sources and nature of this kind of variation, intra-die variations behave in two ways:

- *Pure random*
- *Spatially correlated.*

Pure Random Variations

Pure random variation, also called independent variation, is the device variability that is completely independent of the variation in all other devices, and it does not exhibit spatial correlation or any degree of dependence among them (see Chap. 2). Random dopant fluctuations (RDF) and line edge roughness (LER) are the main contributors of independent variation [28, 40]. With the continuous scaling down of feature sizes and device parameters, the independent intra-die variation component is increasing. Therefore, models such as Pelgrom et al. [47] (which considers the nominal device parameters to quantify the amount of independent variation) and Cao [48] (which also considers short-channel effects) are the best way to account these phenomena.

Figure 3.19 exhibits the behavior of a transistor parameter across a die due to pure random variations. On the top part of the figure, a surface plot of normally distributed random fluctuations of a parameter is illustrated. In this graph, it can be observed that the majority of devices has a value close to the mean (μ) (green and cyan colors) and a few take extreme values (red and blue colors). The random fluctuation of the parameter is better illustrated on the map of gradients at the bottom of the figure. Notice that the values (e.g., green and cyan) around the mean appear randomly at different locations on the die and occupy a large area; and small spots with values (e.g., red and blue) farther from the mean appear randomly on the die.

Spatially Correlated Variations

Spatially correlated variation, also known as a location-dependent variation, is the variability component that exhibits that a parameter deviation of a device is

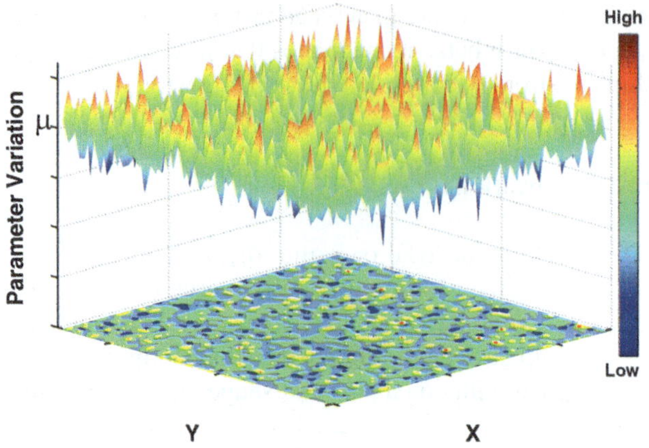

Fig. 3.19 Pure random variations' behavior of a parameter across a die

Fig. 3.20 Spatial correlation behavior between the channel width of two devices

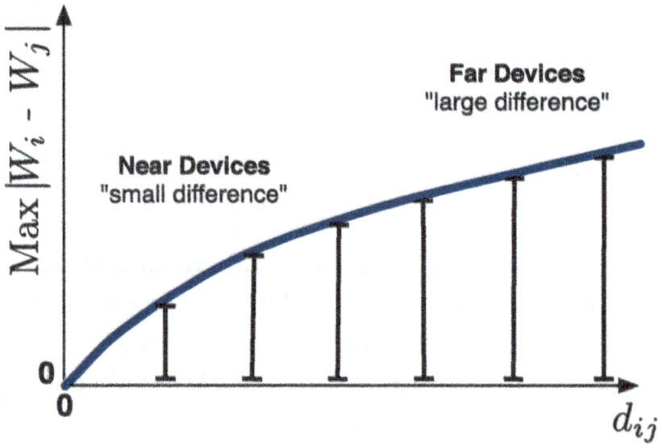

correlated with the deviation of the same parameter of all other devices on the die and the correlation degree between them depends on their physical location on the die. Many of the fundamental processes in the manufacturing flow (such as photolithography, etching, and chemical mechanical polishing) that produce intra-die variation change progressively across the chip [6, 18, 42]. These processes tend to affect near devices in a similar way than those placed far apart between them. This spatial dependence is known as *spatial correlation*. Then, it is because of spatially correlated intra-die variations that devices on the same chip keep a correlation degree among them. Nearby devices on the chip possess a high correlation degree, and this correlation value decreases when the distance between devices increases [18, 49–51]. The main physical parameters that are affected by this variability component are the channel length (L), the channel width (W), and oxide thickness (T_{ox}).

Figure 3.20 plots the variability behavior of the channel width of two transistors as a function of the separation (d_{ij}) between them. Near devices have on average a small difference ($W_i - W_j$) between them, and far devices have on average a large difference ($W_i - W_j$) between them. The solid rising line represents the maximum (Max) difference that can exist between channel widths as the separation between devices increases.

3.5 Parameter Modeling

Statistical device models describing the effect of process variations are required. These models allow to represent the behavior of the parameters varying with process and can be used for understanding circuit behavior with CAD tools. A way to quantify process variation of a parameter is using the linear sum (see Sect. 2.6 of Chap. 2) of the different components of variation. For instance, let X be a normal random variable representing a process parameter. A model of a process parameter considering inter-die and intra-die process variations can be written as follows [50, 52]:

$$X = X_0 + X_{\mathrm{D2D}} + X_{\mathrm{WID}} \tag{3.3}$$

where X_0 represents the nominal value of the parameter. X_{D2D} and X_{WID} are normal variables representing the inter-die and intra-die variations, respectively.

Equation (3.3) is extended as follows:

$$X = X_0 + X_{\mathrm{D2D}} + X_{\mathrm{WID},c} + X_{\mathrm{WID},r} \tag{3.4}$$

where

- Die-to-die variations modeled by X_{D2D} is the component of variation that is shared by all devices in the die.
- Within-die variations modeled by $X_{\mathrm{WID},c}$ is the component of variation specific to each device which is correlated with each other in the die.
- Within-die variations modeled by $X_{\mathrm{WID},r}$ is the component of variation unique to each device which is purely random and does not exhibit any kind of correlation.

Because X is a normally distributed random variable, and variables X_{D2D}, $X_{\mathrm{WID},c}$ and $X_{\mathrm{WID},r}$ are statistically independent between them (see Sect. 2.6 of Chap. 2), the mean and variance of the parameter are determined by

$$\mu_X = X_0 \tag{3.5}$$

$$\sigma_X^2 = \sigma_{X_{\mathrm{D2D}}}^2 + \sigma_{X_{\mathrm{WID},c}}^2 + \sigma_{X_{\mathrm{WID},r}}^2 \tag{3.6}$$

where $\sigma_{X_{\mathrm{D2D}}}^2$ is the variance due to the inter-die variation (i.e., fully correlated in a die), $\sigma_{X_{\mathrm{WID},c}}^2$ is the variance due to the intra-die correlated variation, and $\sigma_{X_{\mathrm{WID},r}}^2$ is the variance due to purely random process variation.

Some device parameters can involve just one component of variation, either correlated or purely random, and other parameters can involve the two components of variations, correlated and purely random. This fact depends on the process steps that define the physical and electrical characteristics of device parameters, such as described in Sects. 3.2 and 3.3.

3.6 Spatial Correlation Modeling

Spatially correlated variation between device parameters must be properly modeled to avoid underestimating the circuit performance. Due to the existence of the correlation between device parameters, the performance characteristics between logic gates (which are composed of devices) on a chip are also spatially correlated. Inter-die and intra-die variations contribute to the total correlation between devices. Different methods have been proposed in the literature for modeling spatial correlation between gates. A summary of the two major models is included below.

3.6.1 Exponential Model

Figure 3.21 shows a schematic of a logic circuit in a chip, where the gates are spatially correlated among them due to process variations. The correlation degree between two gates depends on the separation distance (d_{ij}) between them on the chip.

The degree of spatial correlation between two gates due to their distance separation can be captured using the *exponential model* [51, 53–56]. This model

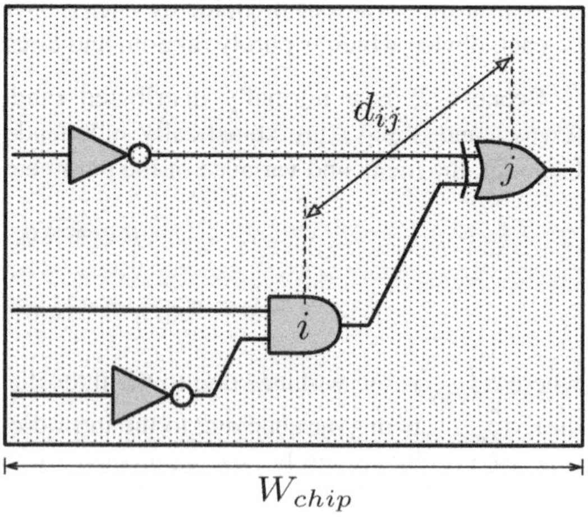

Fig. 3.21 Spatially correlated gates of a digital circuit

assumes that the correlation between two devices decays exponentially as the separation distance between them increases. The empirical and analytical models reported are based on the nature of spatial correlation itself and considering the theory of random fields [57]. The exponential model is the following [51, 54]:

$$\rho(X_i, X_j) = \exp\left(-\frac{d_{ij}}{\mathrm{CD}}\right) \tag{3.7}$$

where

- X_i and X_j represent the parameter X of the gate devices at the location i and j, respectively.
- d_{ij} denotes the separation distance between two gates at positions i and j.
- CD denotes the spatial correlation distance of the considered parameter. It modulates the change rate of the correlation between devices with respect to the distance d_{ij} [55].

It is possible to model inter-die (D2D) and intra-die (WID) spatial correlated variations using (3.7). To accomplish this, two expressions like (3.7) are created, one to model the spatial correlation due to D2D variation and another due to WID variation. The following considerations must be taken into account:

1. The total variation of a parameter is divided into two parts, one part for the inter-die component and the other part for the intra-die one. The percentage assigned to each part can be adjusted according to the used technology node.
2. Two values of spatial correlation distance (CD) are considered, one for inter-die variation and the second for intra-die variation.

According to the defined considerations, the spatial correlation due to die-to-die (D2D) and intra-die variations (WID) for two gates at locations i and j can be estimated as follows:

$$\rho_{\mathrm{D2D}}(X_i, X_j) = K_{\mathrm{D2D}} \exp\left(-\frac{d_{ij}}{\mathrm{CD}_{\mathrm{D2D}}}\right) \tag{3.8}$$

$$\rho_{\mathrm{WID}}(X_i, X_j) = K_{\mathrm{WID}} \exp\left(-\frac{d_{ij}}{\mathrm{CD}_{\mathrm{WID}}}\right) \tag{3.9}$$

where

- $\rho_{\mathrm{D2D}}(X_i, X_j)$ and $\rho_{\mathrm{WID}}(X_i, X_j)$ represent the correlation due to inter-die and intra-die variation, respectively.
- K_{D2D} and K_{WID} are constants that establish the percent of variation related to inter-die and intra-die correlated variations, respectively.
- $\mathrm{CD}_{\mathrm{D2D}}$ and $\mathrm{CD}_{\mathrm{WID}}$ represent the inter-die and intra-die correlation distances, respectively.

The constants K_{D2D} and K_{WID} can be obtained as follows [51, 55]:

$$K_{\text{D2D}} = \frac{\sigma^2_{\text{D2D}}}{\sigma^2_{\text{D2D}} + \sigma^2_{\text{WID},c} + \sigma^2_{\text{WID},r}} \tag{3.10}$$

$$K_{\text{WID}} = \frac{\sigma^2_{\text{WID},c}}{\sigma^2_{\text{D2D}} + \sigma^2_{\text{WID},c} + \sigma^2_{\text{WID},r}} \tag{3.11}$$

where

σ^2_{D2D} represents the inter-die global variation component.
$\sigma^2_{\text{WID},c}$ represents the intra-die spatial variation component.
$\sigma^2_{\text{WID},r}$ represents the intra-die pure random variation component.

The total correlation between gates at locations i and j considering both components of variations, inter-die and intra-die, is obtained by summing (3.8) and (3.9).

$$\rho(X_i, X_j) = \rho_{\text{D2D}}(X_i, X_j) + \rho_{\text{WID}}(X_i, X_j) \tag{3.12}$$

Due to the fact that in Eq. (3.8) the inter-die correlation distance (CD_{D2D}) is much larger than the separation distance between gates (d_{ij}) in the chip, the evaluation of the exponential function of (3.8) results in approximately 1. Thus, (3.8) can be approximated as $\rho(X_i, X_j) = K_{\text{D2D}}$. Through this book, CD_{WID} is calculated with $W_{\text{chip}}/2$ according to [53], but other approximations can be used according to the used technology. The correlation degree between two gates at locations i and j accounting for the inter-die and intra-die variations is modeled as

$$\rho(X_i, X_j) = K_{\text{D2D}} + K_{\text{WID}} \exp\left(-\frac{d_{ij}}{\text{CD}_{\text{WID}}}\right) \tag{3.13}$$

Figure 3.22 illustrates the correlation behavior (ρ) between two gates as a function of the distance (d_{ij}) between them. Three sections can be observed: The first one refers to the inter-die correlated variation; it means that two gates in the same die have a minimum correlation degree independently of the distance between them. This behavior is because of the shared global variation. The second section refers to the intra-die correlated variation behavior as a function of the separation between gates in the die. The correlation value decreases exponentially as the distance increases. The last section refers to pure random variations; it means that the total correlation between nearby gates is always less than one because pure random variation impacts every device in the chip in a different manner.

3.6.1.1 Example

Let us compute the degree of correlation for the gates at positions i and j in the circuit shown in Fig. 3.21. The chip width is $W_{\text{chip}} = 400\,\text{ul}$, and the separation distance

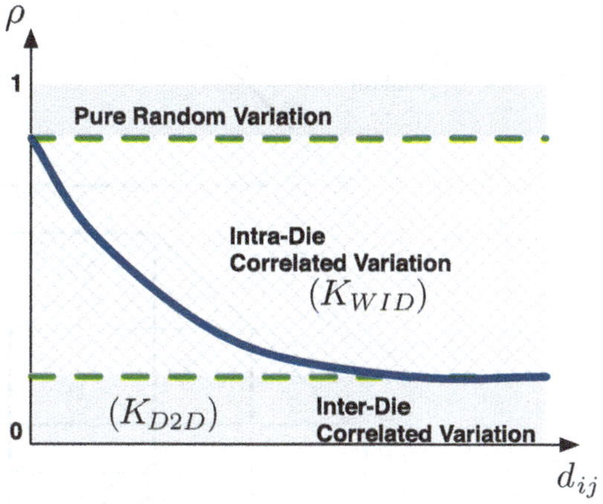

Fig. 3.22 Exponential model for spatially correlated gates. Adapted from [55]

between gates located at position i and j is $d_{ij} = 100\,\text{ul}$. (ul: unit lengths). From the total variation of the parameter, 30% is assumed due to inter-die variation and 70% is assumed due to intra-die variations. The parameter does not have a component of pure random variations ($\sigma^2_{\text{WID},r} = 0$).

The constants that establish the percent of inter-die and intra-die correlated variations result in $K_{\text{D2D}} = 0.3$ and $K_{\text{WID}} = 0.7$ and $\text{CD}_{\text{WID}} = W_{\text{chip}}/2$. Using (3.13), the correlation value between gates located at positions i and j is obtained:

$$\rho = 0.3 + 0.7 \exp\left(-\frac{100}{200}\right) = 0.72$$

Correlations between electrical and physical parameters of the same device also appear [6, 28, 58]. For instance, the value of the threshold voltage (V_{th}) of a MOSFET is correlated with the transistor size (W and L), oxide thickness (T_{ox}), and fluctuations of dopants in the channel [6, 28, 58]. Through this book, only the spatial correlation between parameters of the same type will be considered.

3.6.2 Grid Model

The correlation between gates can also be computed using the grid model such as the quad-tree model proposed by Agarwal et al. [50, 52]. This model allows accounting for the correlation due to the separation distance between gates in the die and also the inter-die correlation. Figure 3.23 exemplifies this model. In this instance, three levels are defined. The first level ($l = 1$) represents the inter-die variations; it consists of just one square because the variation is shared by all devices on the chip. The lower levels ($l = 2$ and $l = 3$) represent the intra-die variation, and each level is hierarchically partitioned. The second level is divided into 4 squares,

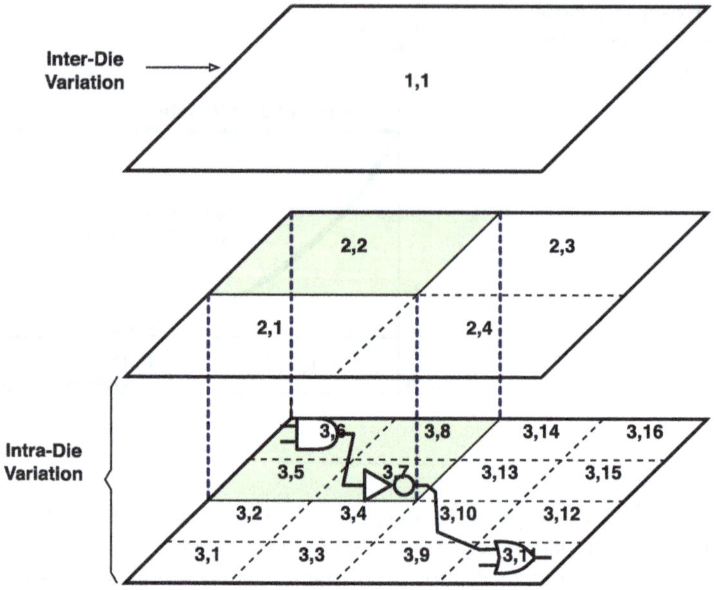

Fig. 3.23 Quad-tree model. Adapted from [50]

and the third level is partitioned into 16 squares. In order to capture the correlation between gates, the total correlated variation of a parameter is divided among all the levels, a percentage is assigned to inter-die variation (first level), and the remaining is destined to intra-die variation (lower levels).

A number of the form $[l, q]$ is assigned to each region, where l represents the level number and q the region number in the grid. Given a parameter X, then a random variable of the form $X_{l,q}$ is created to represent the correlated variation of the parameter in each region $[l, q]$ at all levels. The total correlated variation of the parameter is given by the sum of variables of each region where the device is located, from the bottom to the top level. For the model illustrated in Fig. 3.23, which has three levels, a general expression for the parameter X accounting the inter-die and intra-die correlated variation of a device can be defined as follows:

$$Xc_{3,q} = X_{1,1} + X_{2,q} + X_{3,q} \tag{3.14}$$

where

- $Xc_{3,q}$ represents the total correlated variation of the parameter X.
- $X_{1,1}$ represents the correlated inter-die variation component, which is shared with all devices on the die.
- $X_{2,q}$ and $X_{3,q}$ represent the correlated intra-die variation of the second and third level of the hierarchy, respectively.

In this model, each region of the bottom level can contain many gates. This model can be extended to more levels, such that few gates are placed in each region. The greater number of levels, the higher the accuracy.

Note that this formulation accounts just for the inter-die and intra-die spatial correlated variation. The impact of the intra-die pure random variation component is not quantified in the previous formulation. The pure random variation component of a parameter can be accounted by adapting the quad-tree model as formulated in [42]. Also, the pure random variation component can be accounted during the computation of the gate delays [59, 60].

3.7 Summary

In this chapter, a review of sources of process variations that occur during manufacturing of integrated circuits was presented; also its classification and modeling were given. First, the main processes in the CMOS manufacturing flow have been described. Next, the principal sources of process variations on device parameters and interconnections were given. Photolithography-related processes are the ones that introduce the most variation to the chip. Random dopant fluctuation and line edge roughness are the main contributors of pure random variation. Chemical mechanical polishing is the most important source for interconnecting variations. Then the classification of process parameter variations was given. They can be classified according to how they produce, systematic and nonsystematic. Nonsystematic variations can be classified as inter-die and intra-die variations. The latter can be divided into pure random variation and spatially correlated. A mathematical formulation to model process parameter variations is presented Finally, two models used to quantify the spatial correlation between gates were described; they are the exponential and the quad-tree model.

References

1. J.P. Uyemura, *CMOS Logic Circuit Design* (Kluwer Academic Publishers, New York, 2002)
2. J.M. Rabaey, A. Chandrakasan, B. Nikolić, *Digital Integrated Circuits: A Design Perspective*, 2nd edn. (Prentice Hall, Upper Saddle River, 2003)
3. H. Veendrick, *Nanometer CMOS ICs: From Basics to ASICs* (Springer, New York, 2008)
4. Intel, Moores Law. https://newsroom.intel.com/press-kits/celebrating-the-50th-anniversary-of-moores-law/, January 2016
5. D. Clein, *CMOS IC LAYOUT - Concepts, Methodologies, and Tools*. Technical Contributor: Gregg Shimokura (Newnes, Boston, 2000)
6. M. Orshansky, S.R. Nassif, D. Boning, *Design for Manufacturability and Statistical Design: A Constructive Approach* (Springer, New York, 2008)
7. S. Wolf, *Silicon Processing for the VLSI Era: Deep-Submicron Process Technology* (Lattice Press, Sunset Beach, 2002)
8. M.J. Jackson, *Microfabrication and Nanomanufacturing* (CRC Press, Boca Raton, 2006)

9. Advanced Micro Patterning, http://www.intelligentmp.com/index.html, January 2016
10. The Basics of Microlithography, http://www.lithoguru.com/, January 2016
11. R.F. Pease, S.Y. Chou, Lithography and other patterning techniques for future electronics. Proc. IEEE **96**(2), 248–270 (2008)
12. Applied Materials. http://www.appliedmaterials.com/glossary, January 2016
13. A.C. Jones, M.L. Hitchman, *Chemical Vapour Deposition: Precursors, Processes and Applications* (Royal Society of Chemistry, London, 2009)
14. G. Banerjee, R.L. Rhoades, Chemical mechanical planarization: historical review and future direction. ECS Trans. **13**(4), 1–19 (2008)
15. W. Zhengfeng, Y. Ling, N.S. Huan, T.P. Luan, Chemical Mechanical Planarization. Singapore Institute of Manufacturing Technology, SIMTech Technical Report (PT/01/003/JT), 2001
16. P.B. Zantye, A. Kumara, A.K. Sikderb, Chemical mechanical planarization for microelectronics applications. Mater. Sci. Eng. R. Rep. **45**, 89–220 (2004)
17. M. Wirnshofer, *Variation-Aware Adaptive Voltage Scaling for Digital CMOS Circuits*. Springer Series in Advanced Microelectronics (Springer, New York, 2013)
18. C.C. Chiang, J. Kawa, *Design for Manufacturability and Yield for Nano-Scale CMOS* (Springer, New York, 2007)
19. M. Dietrich, J. Haase, *Process Variations and Probabilistic Integrated Circuit Design* (Springer, New York, 2012)
20. C.A. Mack, *Field Guide to Optical Lithography* (SPIE Press, Bellingham, 2006)
21. L.K. Scheffer, Physical CAD changes to incorporate design for lithography and manufacturability, in *IEEE Design Automation Conference*, 2004, pp. 768–773
22. C. Mack, Understanding focus effects in submicrometer optical lithography: a review. Opt. Eng. **32**(10), 2350–2362 (1993)
23. C. Shin, *Variation-Aware Advanced CMOS Devices and SRAM* (Springer, New York, 2016)
24. C.A. Mack, Optical proximity effects, part 1, 2 and 3. *The Lithography Expert*, FINLE Technologies, 1996
25. P. Bai et al., A 65nm logic technology featuring 35nm gate lengths, enhanced channel strain, 8 cu interconnect layers, low-k ILD and 0.57 μ^2 SRAM cell, in *IEEE International Electron Devices Meeting (IEDM)*, 2004, pp. 657 660
26. C.A. Mack, Fundamental principles of optical lithography: the science of microfabrication (Wiley, New York, 2008)
27. A. Asenov, S. Kaya, A. Brown, Intrinsic parameter fluctuations in decananometer MOSFETs introduced by gate line edge roughness. IEEE J. Solid State Circuits **50**(5), 1254–1260 (2003)
28. J.A. Croon, W. Sansen, H.E. Maes, *Matching Properties of Deep Submicron MOS Transistors* (Springer, New York, 2005)
29. Predictive Technology Model, http://ptm.asu.edu, January 2016
30. H. Huff, D. Gilmer, *High Dielectric Constant Materials: VLSI MOSFET Applications* (Springer, New York, 2005)
31. R.D. Clark, Emerging applications for high K materials in VLSI technology. Materials **7**(4), 2913–2944 (2014)
32. A. Asenov, S. Kaya, J. Davies, Intrinsic threshold voltage fluctuations in decanano MOSFETs due to local oxide thickness variations. IEEE J. Solid-State Circuits **49**(1), 112–119 (2002)
33. K. Takeuchi, A. Nishida, T. Hiramoto, Random fluctuations in scaled MOS devices, in *IEEE Simulation of Semiconductor Processes and Devices (SISPAD)*, 2009
34. Y. Ye, F. Liu, M. Chen, S. Nassif, Y. Cao, Statistical modeling and simulation of threshold variation under random dopant fluctuations and line-edge roughness. IEEE Trans. Very Large Scale Integr. VLSI Syst. **19**, 987–996 (2010)
35. M. Pelgrom, A. Duinmaijer, A. Welbers, Matching properties of MOS transistors. IEEE J. Solid State Circuits **24**(4), 1433–1440 (1989)
36. F. Dekking, C. Kraaikamp, H. Lopuhaa, L. Meester, *A Modern Introduction to Probability and Statistics - Understanding Why and How* (Springer, New York, 2005)
37. D.C. Montgomery, G.C. Runger, *Applied Statistics and Probability for Engineers*, 3rd edn. (Wiley, New York, 2003)

38. G. Nanz, L. Camilletti, Modeling of chemical-mechanical polishing: a review. IEEE Trans. Semicond. Manuf. **8**(4), 382–389 (1995)
39. D. Blaauw, K. Chopra, A. Srivastava, L. Scheffer, Statistical timing analysis: from basic principles to state of the art. IEEE Trans. Comput. Aided Des. Integr. Circuits Syst. **27**(4), 589–607 (2008)
40. S. Bhunia, S. Mukhopadhyay, *Low-Power Variation-Tolerant Design in Nanometer Silicon* (Springer, New York, 2011)
41. A. Chandrakasan, W.J. Bowhill, F. Fox, *Design of High-Performance Microprocessor Circuits* (Wiley, New York, 2000)
42. A. Srivastava, D. Sylvester, D. Blaauw, *Statistical Analysis and Optimisation for VLSI: Timing and Power* (Springer, New York, 2005)
43. C. Forzan, D. Pandini, Statistical static timing analysis: a survey. Integr. VLSI J. **42**(3), 409–435 (2009)
44. S. Reda, S.R. Nassif, Analyzing the impact of process variations on parametric measurements: novel models and applications, in *IEEE Design, Automation and Test in Europe Conference and Exhibition, DATE09*, 2009
45. N.H.E. Weste, D.M. Harris, *CMOS VLSI Design A Circuits and Systems Perspective*, 4th edn. (Addison-Wesley, Boston, 2011)
46. B. Nikoli, L.-T. Pang, Measurements and analysis of process variability in 90nm CMOS, in *8th International Conference on Solid-State and Integrated Circuit Technology*, 2006
47. M. Pelgrom, A. Duinmaijer, A. Welbers, Matching properties of MOS transistors. IEEE J. Solid State Circuits **24**(5), 1433–1439 (1989)
48. Y. Cao, *Predictive Technology Model for Robust Nanoelectronic Design* (Springer, New York, 2011)
49. T. Chawla, Etude de limpact des variations du procede de fabrication sur les circuits numeriques, Ph.D. dissertation, Universidad de Montpellier 2, 2010
50. A. Agarwal, D. Blaauw, V. Zolotov et al., Statistical delay computation considering spatial correlations, in *IEEE ASP-DAC*, January 2003, pp. 271–276
51. J. Xiong, V. Zolotov, L. He, Robust extraction of spatial correlation, in *Proceedings of the 2006 International Symposium on Physical Design, ISPD* (ACM, New York, 2006)
52. A. Agarwal, D. Blaauw, V. Zolotov, Statistical timing analysis for intra-die process variations with spatial correlations, in *IEEE ICCAD*, November 2003, pp. 900–907
53. P. Friedberg, Y. Cao, J. Cain, R. Wang, J. Rabaey, C. Spanos, Modeling within-die spatial correlation effects for process-design co-optimization, in *International Symposium on Quality Electronic Design (ISQED)*, 2005
54. L. Zhang, Y. Hu, C.C.-P. Chen, Statistical timing analysis with path reconvergence and spatial correlations, in *Design, Automation and Test in Europe (DATE)*, 2006
55. J. Xiong, V. Zolotov, L. He, Robust extraction of spatial correlation. IEEE Trans. Comput. Aided Des. Integr. Circuits Syst. **26**(4), 619–631 (2007)
56. N. Lu, Modeling of distance-dependent mismatch and across-chip variations in semiconductor devices. IEEE Trans. Electron Devices **61**(2), 342–350 (2014)
57. E. Vanmarcke, *Random Fields: Analysis and Synthesis* (World Scientific, Singapore, 2010)
58. B.P. Wong, A. Mittal, Y. Cao, G. Starr, *Nano-CMOS Circuit and Physical Design* (Wiley, New York, 2005)
59. K. Kang, B.C. Paul, K. Roy, Statistical timing analysis using levelized covariance propagation, in *IEEE Proceedings of the Design, Automation and Test in Europe Conference and Exhibition (DATE05)*, 2005, pp. 619–631
60. K. Kang, B.C. Paul, K. Roy, Statistical timing analysis using levelized covariance propagation considering systematic and random variations of process parameters. ACM Trans. Des. Autom. Electron. Syst. **11**(4), 848–879 (2006)

Chapter 4
Gate Delay Under Process Variations

Abstract The previous chapter described the nature of the variations of the technological parameters due to the fabrication process. The behavior of correlated and pure random variations was illustrated. This chapter studies the impact of these variations on the statistical performance of logic cells. First, the statistical delay of a logic gate is obtained using simple mathematical formulation. A detailed statistical analysis is devoted to the inverter gate, which due to its simple structure allows an in-depth analysis of its statistical performance. It is shown how to compute the statistical inverter delay. Next, the statistical treatment is extended to a 2-Nand gate. This multi-input gate allows illustrating important related aspects with the serial/parallel connection of transistors that influence the statistical performance. The statistical analysis is generalized to other types of logic gates such as Nor, EX-OR, and AOI/AIO gates. Hand-by-hand examples to compute the delay variance (standard deviation) of logic gates are presented. The behavior of the delay variance of logic gates with relevant circuit parameters is analyzed. Several key design issues affecting the gate delay are illustrated such as sizing a logic gate, load capacitance, input slew time, power supply voltage, and gate delay dependence on input transition.

4.1 Mathematical Formulation of the Statistical Delay of a Logic Gate

Let us represent the statistical delay of a gate with a function D composed of k independent random variables $X_{P_1}, \ldots, X_{P_k}$ (Eq. (4.1)). The random variables correspond to physical and performance parameters. The random variable of a parameter can be approximated as a normal distribution ($X_P \sim N(\mu_{X_P}, \sigma^2_{X_P})$) as shown in Fig. 4.1, where μ_{X_P} and $\sigma^2_{X_P}$ are the mean value and variance of X_P, respectively [1]. The approximation of normal behavior of the parameter distribution can be made even if they are not strictly normal random variables [2, 3]. The foundry usually provides the mean value and the standard deviation of relevant technological and performance parameters.

V. Champac, J. G. Gervacio, *Timing Performance of Nanometer Digital Circuits Under Process Variations*, Frontiers in Electronic Testing 39,
https://doi.org/10.1007/978-3-319-75465-9_4

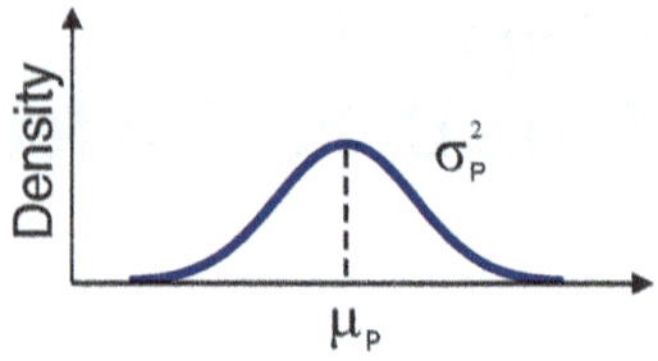

Fig. 4.1 Normal distribution of a technological parameter

$$D = f(X_{P_1}, \ldots, X_{P_k}) \tag{4.1}$$

Complexity to obtain the mean and standard deviation of Eq. (4.1) can be lowered by approximating the function with a Taylor series expansion of first-order [4, 5]. This approximation results in a function D with linear dependence on their parameters (*linear gate delay*) [4–9]. For small variations, the function is expanded around the nominal values of the parameters [10]. The mean (μ_D) and variance (σ_D^2) of the gate delay are estimated in terms of the mean and variance of each parameter.

4.1.1 Mean Delay of a Gate

In Appendix A, a detailed procedure to obtain the variance of a function composed of independent normal random variables is given. A similar procedure can be followed to obtain the mean delay of a function. The generic delay function (Eq. (4.1)) is approximated by a first-order Taylor series expansion. By applying the first moment (expected value) to the expanded delay function and after some operations, a first-order expression of the mean value of the gate delay (μ_D) is obtained [9, 11]:

$$\mu_D = \left[f\left(\mu_{X_{P_1}}, \ldots, \mu_{X_{P_k}}\right) \right]_{\text{nom}} + \frac{1}{2} \sum_{m=1}^{k} \left[\frac{\partial^2 f\left(X_{P_1}, \ldots, X_{P_k}\right)}{\partial X_{P_m}^2} \right]_{\text{nom}} \cdot \sigma_{X_{Pm}}^2 \tag{4.2}$$

where the first term evaluates the function at the nominal values of the parameters and the partial derivatives of the function are evaluated around the nominal values of the parameters.

4.1.2 Variance of the Delay of a Gate

By applying the second moment (variance) to the expanded expression of the delay function and after some operations, a first-order expression of the variance of the gate delay (σ_D^2) is obtained [9, 11]:

$$\sigma_D^2 = \sum_{m=1}^{k} \left[\frac{\partial f\left(X_{P_1}, \ldots, X_{P_k}\right)}{\partial X_{P_m}} \right]_{\text{nom}}^2 \cdot \quad \sigma_{X_{Pm}}^2 \tag{4.3}$$

where the partial derivatives representing the delay sensitivities to small variations of each parameter (S_{D,P_m}) are evaluated at the nominal values of the parameters

$$S_{D,P_m} = \left[\frac{\partial f(X_{P_1}, \ldots, X_{P_k})}{\partial X_{P_m}} \right]_{\text{nom}} \tag{4.4}$$

The variance of the gate delay can be expressed in terms of delay sensitivities as [5, 12, 13]

$$\sigma_D^2 = \sum_{m=1}^{k} S_{D,P_m}^2 \sigma_{X_{P_m}}^2 \tag{4.5}$$

Parameters presenting pure random behavior and spatial correlation can be considered in Eq. (4.5). The mean delay and the sensitivities that are required to estimate the gate delay variance can be obtained using analytical delay equations (e.g., Sakurai delay model [14]) or using SPICE electrical simulator [9, 11]. SPICE (*Simulation Program with Integrated Circuit Emphasis*) is a computer program to simulate semiconductor circuits that was developed at the electronic labs of the University of California at Berkeley [15].

Equations (4.4) and (4.5) show that the variance of the delay is computed in terms of the sum of all the products of the squared delay sensitivities by the variance of each parameter. The contribution of a parameter to the delay variance of the inverter is determined by the relative impact of both the delay sensitivity to a process parameter and the standard deviation of the parameter (Eq. (4.5)) [5, 12, 13].

From Eqs. (4.4) *and* (4.5), *the variance of the gate delay is determined by:*

- The delay sensitivities of a gate to each process parameter (Eqs. (4.4) and (4.5)). The delay variance of a gate increases (decreases) for higher (lower) delay sensitivities.
- The standard deviation of the parameters due to the technological process (Eq. (4.5)).

Let us discuss the validity of the approximation of the gate delay (Eqs. (4.2) and (4.3)) as a normal distribution briefly [12, 16]. In this approximation, the gate delay is represented as a linear function of random variables [17] that represent the process parameters, which are assumed to present normal or Gaussian distribution. A first consideration is that some technological parameters such a the critical dimension (transistor channel length) may present non-normal behavior causing the gate delay deviates from a normal distribution [12, 16, 18–21]. A second consideration is that the gate delay may have a significant nonlinear dependence with a process parameter. Because of this, the gate delay may deviate from a normal distribution as the error due to discarded higher order terms in the linear model becomes significant [12, 16, 18–21]. The error incurred in the linear gate delay model is less significant for intra-chip variations, as they are relatively small in comparison with the nominal values [16]. Non-normal delay distributions could be required for severe parameter variations, leading to a more complex gate delay model including higher-order delay terms [12, 19, 22].

4.2 Delay of Logic Gates Under Process Variations

For nanometer technologies, manufacturing process variations have become quite significant, and they should be included to analyze the behavior of logic gates and circuits [11, 23–27]. Figure 4.2 illustrates the delay behavior of an inverter gate under process variations for a high-to-low input signal. A normal distribution (also called Gaussian distribution) with values of (μ, σ) is applied to the threshold voltage of the pMOS transistor, and the dynamic performance of the inverter is obtained by SPICE Monte Carlo simulations. The output delay curves spread out around the nominal value due to parameter fluctuations as illustrated in Fig. 4.2. The curve located most to the left side corresponds to the fastest response of the inverter delay, and the curve located most to the right side corresponds to the slowest response of the inverter delay. A histogram is built for the low-to-high propagation delay (t_{PLH}) of the inverter from the spreading delay curves. Most of the t_{PLH} values are located around the nominal value, and also higher and lower t_{PLH} values are located farther from the nominal value. The number of cases of t_{PLH} with larger and smaller values than the nominal delay decreases as we move away from the nominal delay. The obtained histogram is approximated with a normal (Gaussian) distribution.

From now on, the focus of this chapter is devoted to the variance (standard deviation) of the delay. The mean of the statistical delay is omitted as it does not depend on process variations and its value can be approximated by the nominal delay.

4.3 Computing Delay Variance of an Inverter

Let us compute the delay variance of the most simple logic structure known as the inverter gate. If the inverter is a simple structure, understanding the delay of this digital gate is quite important because it has properties that are shared with multi-input gates such as Nand/Nor gates and complex gates [28]. The procedure to compute the delay variance of the inverter gate will be extended to the more complex gates. As the CMOS inverter has one single transistor at both its nMOS and pMOS networks (Fig. 4.3), the analysis focuses in the high-to-low propagation time (t_{PHL}). The same methodology also applies for the t_{PLH}.

Parameters W, L, T_{ox}, and V_{th} are the most significant impacting gate delay behavior under process variations [29, 30]. Through this book, parameters W, L, and T_{ox} are used to capture correlated variations (intra-die and inter-die), and V_{th} is

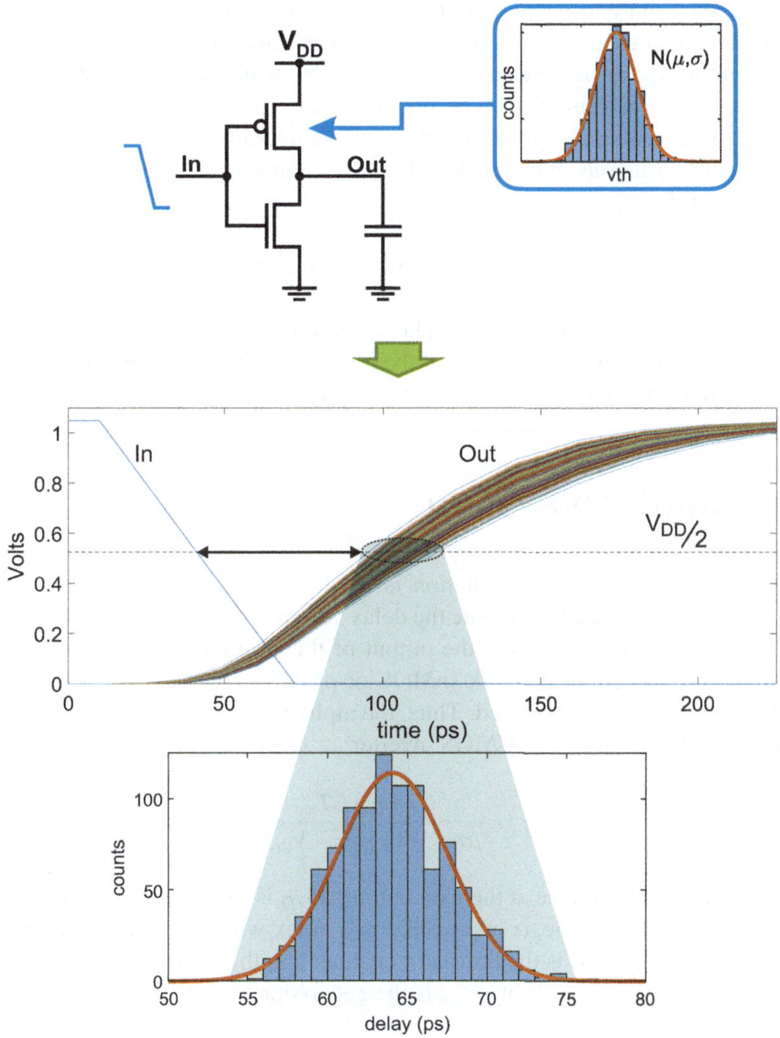

Fig. 4.2 Inverter behavior under process variations

Fig. 4.3 Schematic diagram of a static CMOS inverter

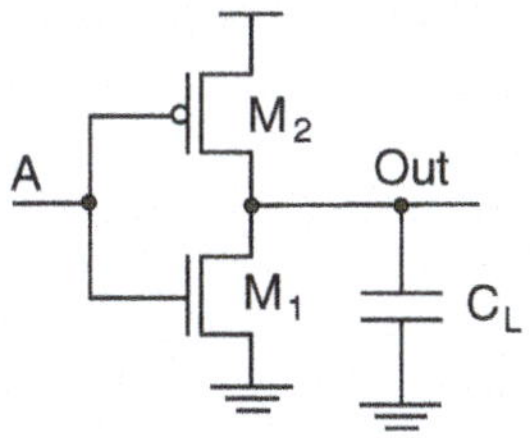

used to capture non-correlated (pure random) variations. However, the same process parameter may present both correlated and pure random variations [6, 31]. The analysis in this section can be easily extended to consider other process parameters. Correlations between different type of parameters [32] are neglected as they are assumed to have a small impact on the gate delay variance. The variance of the gate delay for variations in parameters W, L, T_{ox}, and V_{th} is obtained by expanding Eq. (4.5)) as follows:

$$\sigma_D^2 = S_{D,W}^2\sigma_W^2 + S_{D,L}^2\sigma_L^2 + S_{D,T_{\text{ox}}}^2\sigma_{D,T_{\text{ox}}}^2 + S_{D,V_{\text{th}}}^2\sigma_{V_{\text{th}}}^2 \tag{4.6}$$

First, the variance of the inverter delay is computed using a first-order analytical delay expression. Then, the variance of the inverter delay is computed using the gate delay sensitivity model, which is a more accurate methodology.

4.3.1 Analytical Delay Model

A first-order analytical delay equation is used to obtain simple delay sensitivity expressions that are used to compute the delay variance of the inverter with Eq. (4.6). It is assumed that a transition at the output of the inverter mainly depends on the strength of the switching network (nMOS or pMOS network), which holds while the short-circuit current is limited. Thus, the alpha-power law model [14] is used to obtain the nominal delay of a CMOS inverter:

$$t_D = \frac{C_L V_{DD} L T_{\text{ox}}}{\mu\varepsilon_{\text{ox}} W (V_{DD} - V_{\text{th}})^{\alpha}} \tag{4.7}$$

where C_L is the capacitance at the gate output, V_{DD} is the supply voltage, V_{th} is the transistor threshold voltage, α is the Sakurai's index which can be taken equal to 1 in scaled technologies, L is the transistor channel length, W is the transistor channel width, μ is the carrier's mobility, T_{ox} is the gate oxide thickness, and ε_{ox} is the gate oxide permittivity.

Analytical delay sensitivity expressions are derived from Eq. (4.7), [33]. According to Eq. (4.4), the delay sensitivities of the inverter due to each process parameter are obtained by the partial derivative of Eq. (4.7) with respect to each process parameter. For the high-to-low propagation delay, the delay sensitivities are

$$S_{D,W} = -\frac{C_L V_{DD}}{\mu\varepsilon_{\text{ox}} (V_{DD} - V_{\text{th}})^{\alpha}} \cdot \frac{T_{\text{ox}} L}{W^2} \tag{4.8a}$$

$$S_{D,L} = \frac{C_L V_{DD}}{\mu\varepsilon_{\text{ox}} (V_{DD} - V_{\text{th}})^{\alpha}} \cdot \frac{T_{\text{ox}}}{W} \tag{4.8b}$$

$$S_{D,T_{\text{ox}}} = \frac{C_L V_{DD}}{\mu\varepsilon_{\text{ox}} (V_{DD} - V_{\text{th}})^{\alpha}} \cdot \frac{L}{W} \tag{4.8c}$$

$$S_{D,V_{\text{th}}} = \frac{\alpha C_L V_{DD}}{\mu \varepsilon_{\text{ox}} (V_{DD} - V_{\text{th}})^{\alpha+1}} \cdot \frac{T_{\text{ox}} L}{W} \tag{4.8d}$$

The delay sensitivities and hence the gate delay variance (Eq. (4.6)*) depend on the following circuit design parameters:*

- Transistor driving strength (channel width W).
- Load capacitance (C_L).
- Supply voltage.
- Input slew time.

Equations (4.8a)–(4.8d) provide important insights into the impact of some circuit design parameters on the variance of the gate delay. Equations (4.8a)–(4.8d) show explicitly the dependence of the delay sensitivities with W and C_L. Delay sensitivities reduce as the transistor driving strength increases, and they increase as the load capacitance increases. Besides, the delay sensitivities also depend on the supply voltage as can be observed in Eq. (4.8). The circuit behavior with V_{DD} is a concern for applications that require to lower V_{DD} and for multi-V_{DD} design. The dependence of the delay sensitivities with the input transition time is not appreciated in Eq. (4.8) due to the used first-order gate delay expression.

Estimation of the gate delay variance using Eqs. (4.8a)–(4.8d) neglects the impact of the short-circuit current and short-channel effects. More accurate gate delay expression can be used but the complexity of the expressions increases.

4.3.2 Sensitivity Delay Model

The inverter delay variance computed using analytical expressions provides valuable information about inverter behavior in the presence of process variations. However, inaccuracies appear in scaled technologies due to more prominent short-channel effects [24, 34]. Short-channel effects are difficult to be considered in first-order analytical expressions. Furthermore, body effect in serially connected transistors when using Nand and Nor type gates makes this a more difficult problem. A more accurate computation of the inverter delay under process variations uses pre-computed gate delay sensitivities obtained doing transient simulations with an electrical simulator as SPICE [4, 5, 7, 35, 36]. This methodology allows capturing the impact of input slew time, nonlinear gate capacitances, multi-input gate structure, and higher-order current behavior. This methodology is mainly used from now on unless otherwise specified. Sensitivity-based delay models are accurate as long as parameter variation is small [37]. The delay sensitivity of a function is the ratio of change of the delay to a change in a single parameter X_{P_m} [33, 38]. The delay sensitivity is defined as

$$S_{D,P_m} = \left[\frac{\partial t_D}{\partial X_{P_m}} \right]_{\text{nom}} \tag{4.9}$$

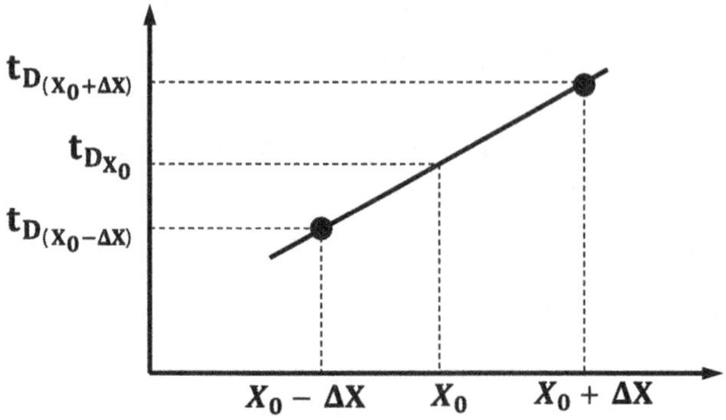

Fig. 4.4 Illustrative example of sensitivity computation

The calculation of the delay sensitivity to a parameter is illustrated in Fig. 4.4. First, the gate delay is measured for a transient simulation made with the process parameter of interest at a ΔX lower than its nominal value with the other parameters held at their nominal values. Second, the gate delay is measured for a transient simulation made with the process parameter of interest at a ΔX higher than its nominal value with the other parameters held at their nominal values. The delay sensitivity to a single parameter is calculated with the following expression (Fig. 4.4) [38]:

$$S_{D,P_m} \approx \frac{t_{D(X_0+\Delta X)} - t_{D(X_0-\Delta X)}}{(X_0 + \Delta X) - (X_0 - \Delta X)} \tag{4.10}$$

In SPICE, $X_0+\sigma_X$ and $X_0-\sigma_X$ are used for $X_0+\Delta X$ and $X_0-\Delta X$, respectively. Once all the delay sensitivities to each process parameter variation are obtained, they are used in Eq. (4.6) to compute the gate delay variance. The delay sensitivities may be pre-computed for each gate of a digital library [36, 38]. They may be obtained as a function of (1) load capacitance, (2) combinations of input transition(s), and (3) input transition time. This requires some computational effort, but it is only made once for a digital library of the technology. More readings material about the estimation of the delay sensitivity can be found in [24, 39].

4.3.3 *Example of Computing Delay Standard Deviation of an Inverter*

The computation of the standard deviation of the delay of an inverter using the sensitivity delay model is presented. The standard deviation of the delay is also obtained with a SPICE Monte Carlo simulation. A minimum-sized inverter with $W_N = 120\,\text{nm}$, $W_P = 240\,\text{nm}$, and minimum channel length is designed. The

Table 4.1 Standard deviation of technological parameters

Standard deviation	Nmos	Pmos
σ_W (nm)	6	6
σ_L (nm)	3	3
$\sigma_{T_{ox}}$ (nm)	0.119	0.121
$\sigma_{V_{th}}$ (mV)	36.7	35.8

Value of $\sigma_{V_{th}}$ does not account for RDF effect

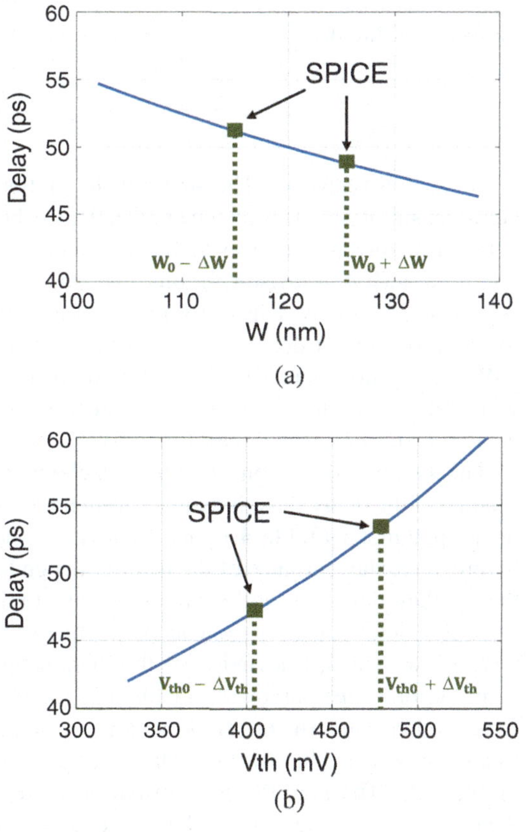

Fig. 4.5 Estimation of the delay sensitivities using SPICE for the high-to-low propagation delay. (**a**) Transistor channel width. (**b**) Threshold voltage

inverter has similar high-to-low (t_{PHL}) and low-to-high (t_{PLH}) propagation delays. The inverter drives a load capacitance of eight minimum-sized inverters. Table 4.1 shows the standard deviations for the parameters W, L, T_{ox}, and V_{th}. The value of the standard deviation of the threshold voltage shown in Table 4.1 corresponds to a transistor with minimum dimensions. The transistor threshold voltage accounting the effect of RDF is calculated using Eq. (3.2).

The delay sensitivities are estimated using SPICE electrical simulator for both input transitions. First, the propagation delays are measured for two values of W and also two values of V_{th}. Figure 4.5a plots t_{PHL} versus the transistor channel width (W)

Table 4.2 Delay sensitivities of the inverter for t_{PHL} and t_{PLH}

Delay sensitivity	t_{PHL}	t_{PLH}
$S_{D,W}$ (ps/nm)	−0.200	−0.109
$S_{D,L}$ (ps/nm)	0.215	0.242
$S_{D,T_{\mathrm{ox}}}$ (ps/nm)	4.922	5.588
$S_{D,V_{\mathrm{th}}}$ (ps/mV)	0.0563	0.0576

Table 4.3 Variance contributions of each process parameter to the total variance of the inverter delay

Variance (ps^2)	t_{PHL}	t_{PLH}
$(S_{D,W}\sigma_W)^2$	1.44	0.427
$(S_{D,L}\sigma_L)^2$	0.416	0.527
$(S_{D,T_{\mathrm{ox}}}\sigma_{T_{\mathrm{ox}}})^2$	0.343	0.457
$(S_{D,V_{\mathrm{th}}}\sigma_{V_{\mathrm{th}}})^2$	4.269	2.126

of the nMOS transistor. The slope of the plotted points is negative, and hence the delay sensitivity to W is also negative. This is because the high-to-low propagation delay of the inverter reduces as W_N increases. Similarly, Fig. 4.5b plots t_{PHL} versus V_{th} of the nMOS transistor. In this case, the slope of the plotted points is positive, and hence the delay sensitivity to V_{th} is also positive. This is because the high-to-low propagation delay of the inverter increases as the threshold voltage of the nMOS transistor increases. Table 4.2 summarizes the estimated delay sensitivities with SPICE data for the considered process parameters. Note that the sign of the delay sensitivities agrees with Eq. (4.8).

The delay variance contribution of each parameter (Table 4.3) is obtained by the product of the delay sensitivity to a parameter (Table 4.2) with the standard deviation of the parameter (Table 4.1) and then rising the product to the square. As stated before, the delay variance of the inverter depends on both the delay sensitivities and the standard deviation of the parameters due to the technological process.

A high delay sensitivity to gate oxide thickness fluctuations (Table 4.2) combined with a low standard deviation of this parameter (Table 4.1) results in a moderate contribution of the parameter fluctuations to the inverter delay variance (Table 4.3). On the other hand, the threshold voltage is the process parameter most contributing to the variance of the inverter delay in spite that it presents a low delay sensitivity (Table 4.2). The significant contribution of the threshold voltage to the inverter delay variance is due to its higher standard deviation due to the fabrication process (Table 4.1).

The variance of t_{PHL} is obtained by summing up each one of the delay variance contributions due to correlated parameters and due to pure random parameter variations (Eq. (4.6)) given in Table 4.3. A similar procedure is made for the t_{PLH}. The standard deviation of t_{PHL} is obtained by the square root of the variance. The standard deviations of t_{PHL} and t_{PLH} are 2.553 and 1.883 ps, respectively. The standard deviation of the delay of an inverter also is obtained using a SPICE Monte Carlo simulation. The dependence of the standard deviation of V_{th} on the transistor channel width due to RDF is properly accounted. The standard deviation of the delay of the inverter obtained using the sensitivity delay model and also using SPICE is

Table 4.4 Standard deviation of the delay of the inverter

Standard deviation (ps)	t_{PHL}	t_{PLH}
σ_D (Sensitivity delay model)	2.543	1.881
σ_D (SPICE Monte Carlo)	2.568	2.001

given in Table 4.4. It is observed that both estimations are very close. The error of the standard deviation is −0.626% and −5.819% for t_{PHL} and t_{PLH}, respectively. We estimated the delay sensitivity to a parameter with only two points (Fig. 4.5). Using more points to estimate the delay sensitivity reduces the error.

4.4 Computing Delay Variance of a Nand Gate

Multiple-input logic gates (e.g., Nand, Nor, AOI, and AIO gates) present more complex structures in the nMOS and pMOS networks, which modify the statistical behavior. First, the high-to-low propagation delay of a Nand gate, on which the output transition takes place through serially connected nMOS transistors, is analyzed. Second, the low-to-high propagation delay of a Nand gate, on which the output transition takes place through parallel connected pMOS transistors, is analyzed. The extension of this analysis to other multiple-input logic gates is straightforward.

4.4.1 Falling Transition of a Nand Gate

Let us analyze the high-to-low propagation delay (t_{PHL}) of a Nand gate. The serially connected transistors in the nMOS network can present both pure random and correlated variations. W, L, and T_{ox} are assumed to present only correlated variations, and the transistor threshold voltage (V_{th}) is assumed to present only pure random variations. The variance of the delay of a Nand gate due to both correlated and independent parameters can be obtained with a similar procedure to that followed to obtain the variance of the delay of an inverter. In a similar way to the inverter case, the analysis of the variance of the t_{PHL} in a 2-Nand gate neglects fluctuations in the pMOS network as they are assumed to have a minor impact on the delay variance.

4.4.1.1 Nand Delay due to Correlated Parameters with General Variance Model

The variance of the Nand delay due to correlated parameters can be approximated by a first-order Taylor series expansion of the generic gate delay function and then using some statistical properties such as shown in Appendix A. For simplicity of the

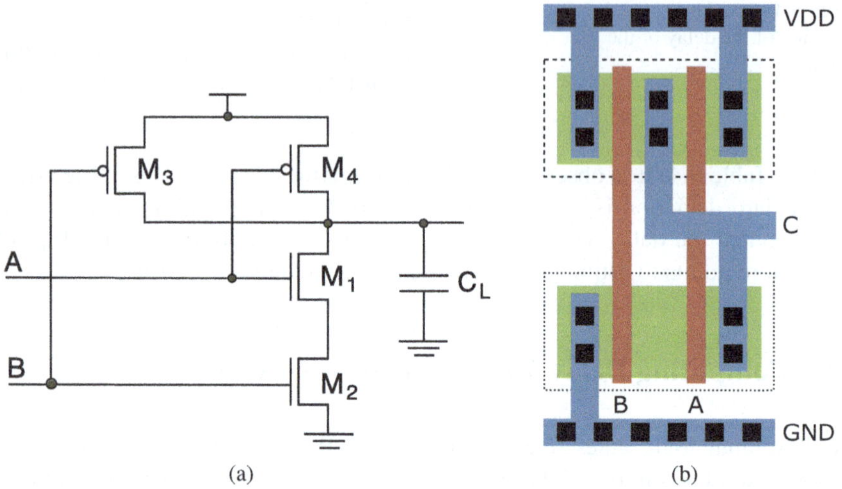

Fig. 4.6 2-Nand gate. (**a**) Schematic circuit. (**b**) Layout

analysis, only correlated variations in the channel width of the nMOS transistors M_1 and M_2 of a 2-Nand gate (Fig. 4.6a) are considered. Then, the analysis is extended to other correlated parameters in the 2-Nand gate. Thus, from Eq. (A.8) in Appendix A, the variance of the delay of the 2-Nand gate is given by

$$\sigma_{D,W}^2 = \left[\frac{\partial f\left(X_{P_1},\ldots,X_{P_k}\right)}{\partial X_{W1}}\right]_{\text{nom}}^2 \sigma_{W1}^2 + \left[\frac{\partial f\left(X_{P_1},\ldots,X_{P_k}\right)}{\partial X_{W2}}\right]_{\text{nom}}^2 \sigma_{W2}^2 + 2\left[\frac{\partial f\left(X_{P_1},\ldots,X_{P_k}\right)}{\partial X_{W1}}\right]_{\text{nom}} \left[\frac{\partial f\left(X_{P_1},\ldots,X_{P_k}\right)}{\partial X_{W2}}\right]_{\text{nom}} \rho_{X_{W1},X_{W2}}\sigma_{W1}\sigma_{W2} \tag{4.11}$$

where $\rho_{X_{W1,W2}}$ is the correlation between the distributions of the parameters W_1 and W_2, σ_{W1} is the standard deviation of W_1, and σ_{W2} is the standard deviation of W_2. The first term is the variance of the Nand delay due to variations only in W_1, the second term is the variance of the Nand delay due to variations only in W_2, and the third term is the covariance between the delay distributions due to W_1 and W_2.

Since the transistors in the layout of a small cell (Fig. 4.6b) are usually very close, it can be assumed that the correlated variations impact similarly to the devices of a Nand gate [40, 41]. As a consequence, the distributions of $W1$ and $W2$ can be assumed fully correlated ($\rho_{X_{W1},X_{W2}} = 1$). Replacing this condition in Eq. (4.11) and using delay sensitivities, the variance of the Nand delay due to the correlated parameters W_1 and W_2 is written as

$$\sigma_{D,W}^2 = S_{D,W1}^2\sigma_{W1}^2 + S_{D,W2}^2\sigma_{W2}^2 + 2S_{D,W1}S_{D,W2}\sigma_{W1}\sigma_{W2} \tag{4.12}$$

where $S_{D,W1}$ and $S_{D,W2}$ are the delay sensitivities to small variations only in W_1 and only in W_2, respectively. It should be noted that σ_{W1} and σ_{W2} have the same value because they refer to the same parameter.

A similar expression to Eq. (4.12) can be obtained for other correlated parameters (e.g., L, T_{ox}).

The variance of the delay of a 2-Nand gate due to all the correlated parameters can be computed by

$$\begin{aligned}\sigma^2_{D_{\text{corr}}} &= S^2_{D,W1}\sigma^2_{W1} + S^2_{D,W2}\sigma^2_{W2} + 2S_{D,W1}S_{D,W2}\sigma_{W1}\sigma_{W2}\\ &+ S^2_{D,L1}\sigma^2_{L1} + S^2_{D,L2}\sigma^2_{L2} + 2S_{D,L1}S_{D,L2}\sigma_{L1}\sigma_{L2}\\ &+ S^2_{D,T\text{ox}1}\sigma^2_{T\text{ox}1} + S^2_{D,T\text{ox}2}\sigma^2_{T\text{ox}2} + 2S_{D,T\text{ox}1}S_{D,T\text{ox}2}\sigma_{T\text{ox}1}\sigma_{T\text{ox}2}\end{aligned} \tag{4.13}$$

where the delay sensitivities are obtained making variations only in one transistor at a time such as in Eq. (4.12).

4.4.1.2 Nand Delay due to Correlated Parameters with a Compact Model

The computation of the delay variance due to correlated parameters using Eq. (4.13) requires the estimation of two delay sensitivities for each considered correlated parameter. A compact computation of the delay variance can be used to reduce the number of terms of delay sensitivities. The basis of this model can be derived from Eq. (4.12). Let us use σ_W for both σ_{W1} and σ_{W2} because they refer to the process fluctuation of the same process parameter. Using this consideration, Eq. (4.12) can be rewritten as follows:

$$\sigma^2_{D,W} = (S^2_{D,W1} + S^2_{D,W2} + 2S_{D,W1}S_{D,W2})\sigma^2_W = (S_{D,W1} + S_{D,W2})^2\sigma^2_W \tag{4.14}$$

Assuming linearity ($S_{D,W1}+S_{D,W2} = S_{D,W}$), the previous expression reduces to

$$\sigma^2_{D,W} = S^2_{D,W}\sigma^2_W \tag{4.15}$$

where now the delay sensitivity to small variations in channel width ($S_{D,W}$) is obtained by making simultaneously the same parameter variation at both nMOS transistors of a 2-Nand gate. In other words, the same parameter variable is assigned to both transistors to estimate the delay sensitivity.

The variance of the delay of a 2-Nand gate due to all the correlated parameters can be computed by

$$\sigma^2_{D_{\text{corr}}} = S^2_{D,W}\sigma^2_W + S^2_{D,L}\sigma^2_L + S^2_{D,T_{\text{ox}}}\sigma^2_{T_{\text{ox}}} \tag{4.16}$$

where the delay sensitivities are obtained by making simultaneously the same parameter variation at both nMOS transistors of a 2-Nand gate. In this case, three delay sensitivities are estimated instead of six delay sensitivities required in Eq. (4.13).

4.4.1.3 Nand Delay due to a Parameter with Pure Random Variations

Let us assume that the transistor threshold voltage is affected only by pure random variations. In this case, the same process parameter may have different values in spite of the proximity between devices. The correlation between the distributions of parameters V_{th1} and V_{th2} of the nMOS transistors M1 and M2 of a 2-Nand gate (Fig. 4.6a) can be considered zero. Replacing this condition in Eq. (A.8) of Appendix A, the delay variance of a 2-Nand gate due to pure random variations in the threshold voltage is given by

$$\sigma^2_{D\text{ind}} = \left[\frac{\partial f\left(X_{P_1}, \ldots, X_{P_k}\right)}{\partial X_{V_{\text{th1}}}}\right]^2_{\text{nom}} \sigma^2_{V_{\text{th1}}} + \left[\frac{\partial f\left(X_{P_1}, \ldots, X_{P_k}\right)}{\partial X_{V_{\text{th2}}}}\right]^2_{\text{nom}} \sigma^2_{V_{\text{th2}}} \tag{4.17}$$

The previous expression is rewritten in terms of delay sensitivities as

$$\sigma^2_{D\text{ind}} = S^2_{D,V_{\text{th1}}} \sigma^2_{V_{\text{th1}}} + S^2_{D,V_{\text{th2}}} \sigma^2_{V_{\text{th2}}} \tag{4.18}$$

where $S_{D,V_{\text{th1}}}$ is the delay sensitivity to small variations only in V_{th1} and $S_{D,V_{\text{th2}}}$ is the delay sensitivity to small variations only in V_{th2}. It should be noted that $\sigma_{V_{\text{th1}}}$ and $\sigma_{V_{\text{th2}}}$ have the same value because they refer to the same parameter.

4.4.1.4 Total Delay Variance of a Nand Gate

The total delay variance of a Nand gate is the sum of the variance contributions due to correlated parameters and due to pure random parameter variations. The total delay variance of the 2-Nand gate using the compact model for the correlated parameters is given by

$$\sigma^2_D = S^2_{D,W}\sigma^2_W + S^2_{D,L}\sigma^2_L + S^2_{D,T_{\text{ox}}}\sigma^2_{T_{\text{ox}}} + +S^2_{D,V_{\text{th1}}}\sigma^2_{V_{\text{th1}}} + S^2_{D,V_{\text{th2}}}\sigma^2_{V_{\text{th2}}} \tag{4.19}$$

where the delay sensitivities for spatially correlated parameters are estimated with the compact model and the delay sensitivities for independent parameters are estimated separately for each transistor. The delay variance contribution of each parameter is appreciated in the previous equation.

4.4.2 Dependence of Nand Delay on Input Transition

Another important issue with multi-input gates is the fact that a transition at the output of the gate could occur due to different input switching combinations [28, 42–44]. Hence, the delay sensitivities should be evaluated for each of the possible input

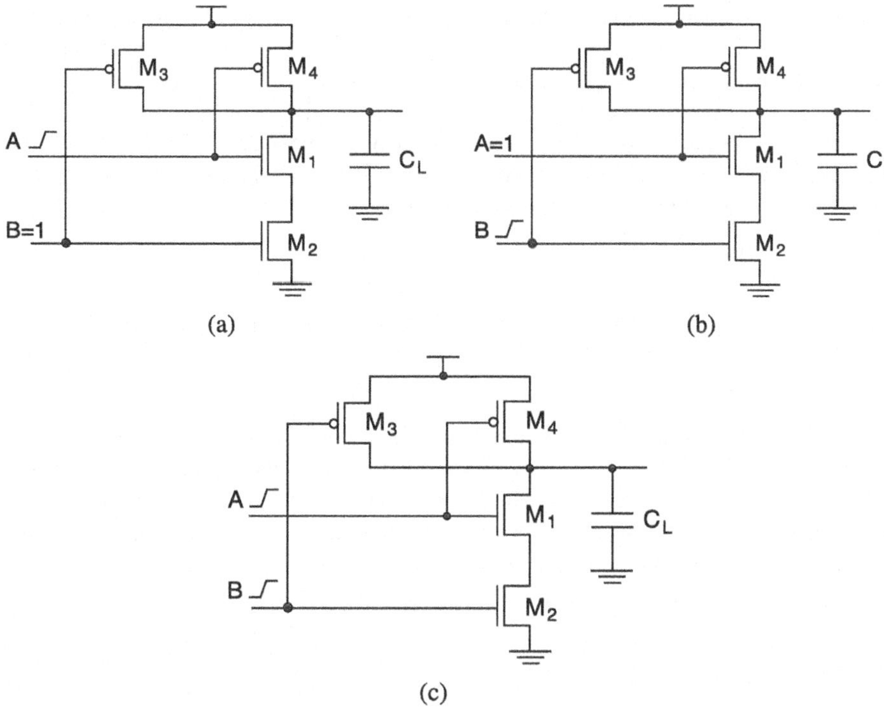

Fig. 4.7 Conditions to compute delay sensitivities of t_{PHL} in a 2-Nand gate for different combinations of input transitions. (**a**) Single transition at input A. (**b**) Single transition at input B. (**c**) Simultaneous transition at inputs A and B

transitions causing the same output transition. Figure 4.7 shows the conditions to compute the delay sensitivities of the high-to-low propagation delay for different combinations of input transitions. Figure 4.7a gives the input conditions in a 2-Nand gate to compute the delay sensitivities for a single transition at input A, Fig. 4.7b for a single transition at input B, and Fig. 4.7c for simultaneous input transition at inputs A and B. Using these input conditions, the delay sensitivity to a small variation of a parameter can be estimated using an electrical simulator as SPICE [9, 11].

On the other hand, there are three input combinations than could produce a low-to-high output transition in a 2-Nand gate [42]: (1) $A = 1$ and $B = 1 \rightarrow 0$, $A = 1 \rightarrow 0$ and $B = 1$, and $A = 1 \rightarrow 0$ and $B = 1 \rightarrow 0$.

4.4.3 Rising Transition of a Nand Gate

The variance of the low-to-high propagation delay (t_{PLH}) in a 2-Nand gate can be computed using the previously presented methodologies. Single and simultaneous input transitions may turn on one and two pMOS transistors, respectively. For a single input transition, the variance of the delay of a 2-Nand gate is computed similarly to the inverter gate as only one pMOS transistor is turned on. For simultaneous switching at the inputs, the variance of the t_{PLH} in a 2-Nand gate is computed with the methodology previously presented for the t_{PHL} of the 2-Nand gate with (1) delay sensitivities of correlated parameters are estimated using the compact model and (2) delay sensitivities of parameters with pure random variations are estimated separately for each transistor.

4.4.4 Example of Computing the Delay Variance of a Nand Gate

This example illustrates the procedure to estimate the variance of the high-to-low (t_{PHL}) and low-to-high (t_{PLH}) propagation delays of a 2-Nand gate. A symmetrical 2-Nand gate ($W_N = 240\,\text{nm}$, $WP = 240\,\text{nm}$) with delays similar to the minimum-sized inverter is designed. The Nand drives a load capacitance of eight minimum-sized inverters. The standard deviations of the parameters W, L, T_{ox}, and V_{th} are those given in Table 4.1. The transistor threshold voltages accounting the effect of RDF are calculated using Eq. (3.2). The variance of the t_{PHL} is computed for a single transition at input A (Fig. 4.7a). The variance of t_{PLH} is computed for a single transition at input B of a 2-Nand gate. In this case, A input is set to 1 logic and B makes a $1 \rightarrow 0$ transition (Fig. 4.7). The variance of the delay is estimated considering variations only in those transistors of the network making the transition.

For the *delay sensitivity model* (Eq. (4.19)), the delay sensitivities of correlated parameters are estimated using the compact model, and the delay sensitivities of a parameter with pure random variations are estimated separately for each transistor. Table 4.5 shows the estimated delay sensitivities with SPICE for both input transitions.

Table 4.6 shows the delay variance contribution of each parameter. This is obtained by the product of the delay sensitivity to a parameter (Table 4.5) with the standard deviation of the parameter (Table 4.1) and then rising the product to the square.

The variance of t_{PHL} is obtained by summing up each one of the delay variance contributions due to correlated parameters and due to pure random parameter

Table 4.5 Delay sensitivities of the 2-Nand gate

Delay sensitivity	t_{PHL}	t_{PLH}
$S_{D,W}$ (ps/nm)	−0.119	−0.093
$S_{D,L}$ (ps/nm)	0.370	0.215
$S_{D,T_{ox}}$ (ps/nm)	5.452	4.748
$S_{D,V_{th1}}$ (ps/mV)	0.068	–
$S_{D,V_{th2}}$ (ps/mV)	0.023	–
$S_{D,V_{th3}}$ (ps/mV)	–	0.050

Table 4.6 Variance contributions of each process parameter to the total variance of the 2-Nand gate

Variance (ps^2)	t_{PHL}	t_{PLH}
$(S_{D,W}\sigma_W)^2$	0.510	0.311
$(S_{D,L}\sigma_L)^2$	1.232	0.416
$(S_{D,T_{ox}}\sigma_{T_{ox}})^2$	0.421	0.330
$(S_{D,V_{th1}}\sigma_{V_{th1}})^2$	3.114	–
$(S_{D,V_{th2}}\sigma_{V_{th2}})^2$	0.356	–
$(S_{D,V_{th3}}\sigma_{V_{th3}})^2$	–	1.602

Table 4.7 Standard deviation of the delay of the 2-Nand gate

Standard deviation (ps)	t_{PHL}	t_{PLH}
σ_D (delay sensitivity model)	2.373	1.63
σ_D (SPICE Monte Carlo)	2.369	1.584

variations (Eq. (4.19)) given in Table 4.6. A similar procedure is made for the t_{PLH}. The square root of the delay variance is made to obtain the standard deviation. The standard deviations of t_{PHL} and t_{PLH} are 2.382 and 1.634 ps, respectively.

The standard deviation of the delay of the 2-Nand gate is also obtained with a SPICE Monte Carlo simulation. Table 4.7 shows the standard deviation using the delay sensitivity model and SPICE. The error of the computed standard deviation is 0.546% and 3.060% for t_{PHL} and t_{PLH}, respectively. We estimated the delay sensitivity to a parameter with only two points (Fig. 4.5). Using more points to estimate the delay sensitivity reduces the error.

4.5 Computing Statistical Delay of a General Logic Gate

The mean delay of any gate can be obtained by evaluating the delay function of the gate at the nominal values of the parameters. This can be approximated by

$$\mu_D = \left[f\left(\mu_{X_{P_1}}, \ldots, \mu_{X_{P_k}}\right) \right]_{\text{nom}} \tag{4.20}$$

For a better accuracy, the mean delay of a gate is evaluated using SPICE electrical simulator.

The analysis presented for the inverter and Nand gates can be extended to other more complex gates. The analysis made for serial (parallel) connected transistors in the Nand gate can be applied to other gates having serial (parallel) connected transistors in their networks. In a Nor gate, the variance of the delay due to switching of the pMOS network (serial-connected transistors) can be treated similarly to the nMOS network (serial-connected transistors) of the Nand gate. On the other hand, the variance of the delay due to switching of the nMOS network (parallel-connected transistors) in a Nor gate can be treated similarly to the pMOS network (parallel-connected transistors) of the Nand gate. Similarly, the statistical performance of other types of gates (e.g., EX-OR and AOI/AIO gates) can also be analyzed.

The variance of the delay of any gate considering both parameters with pure random variations and correlated parameters (σ_D^2) is obtained by

$$\sigma_D^2 = S_{D,W}^2\sigma_W^2 + S_{D,L}^2\sigma_L^2 + S_{D,T_{\text{ox}}}^2\sigma_{T_{\text{ox}}}^2 + \sum_{i=1}^{p} S_{D,V\text{th}_i}\sigma_{V\text{th}_i} \tag{4.21}$$

where $S_{D,W}$, $S_{D,L}$, and $S_{D,T_{\text{ox}}}$ are the delay sensitivities due to correlated parameters for the input transitions of interest. The delay sensitivities are obtained with the compact model. For multi-input gates, the delay sensitivities are estimated considering variations at the transistors of the switching network that is activated by the input signal transitions. On the other hand, $S_{D,V_{\text{th}_i}}$ is the delay sensitivity to V_{th} of transistor i for the input transition of interest. The delay sensitivities of independent parameters are computed separately for each activated transistor of the activated network by the input signal transitions.

4.6 Design Issues Impacting Gate Delay Under Process Variations

In this section, the impact of the main circuit parameters, which are controllable by the designer in some way, on the standard deviation of the inverter delay is analyzed. The considered parameters are the transistor channel width, load capacitance, and input transition time. Besides, the impact of power supply voltage is also analyzed. The values of these parameters determine the mean and standard deviation of the gate delay. The focus of the analysis is devoted to the standard deviation of the gate delay. In [45], the impact of the parameters on the delay variability (σ/μ) of logic gates is studied.

4.6.1 Sizing a Logic Gate

Equation (4.8) shows that the delay sensitivities, and hence the standard deviation of the inverter delay, strongly depend on the transistor channel width, which is a circuit parameter controllable by a designer. Equation (4.8a) shows that the delay sensitivity to small variations of W decreases quadratically as the transistor channel width increases. Furthermore, an increment of W also reduces the other delay sensitivities (Eqs. (4.8b)–(4.8d)). Figure 4.8 shows that the standard deviation of the inverter delay decreases as the inverter is sized up [26, 46]. 1× corresponds to the minimum-sized symmetrical inverter. In this inverter, the nMOS transistor channel width is designed with the minimum size allowed by the technology, and the pMOS transistor channel is sized up to have symmetrical behavior. Two clear operation regions appear. In the first region (left side in Fig. 4.8), the delay standard deviation decreases significantly as the inverter size increases. In the second region (right side

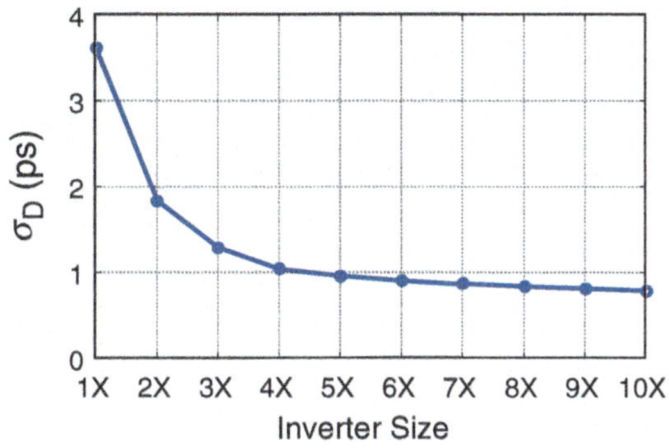

Fig. 4.8 Influence of transistor sizing

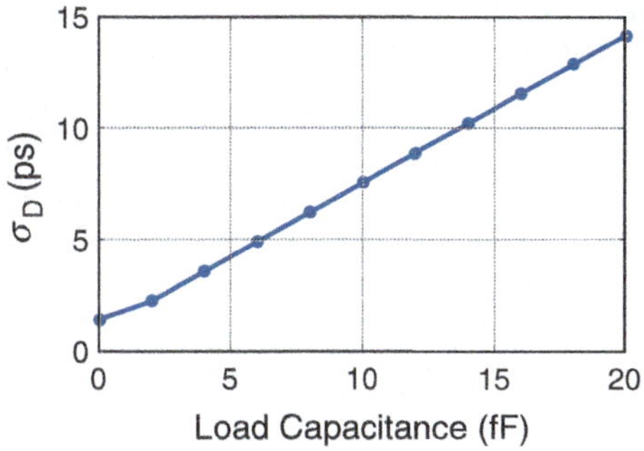

Fig. 4.9 Influence of load capacitance

in Fig. 4.8), the delay standard deviation presents only small reductions of the delay standard deviation as the inverter sizes further increase. It must be noticed that the self-loading capacitance of the inverter increases as it is sized up. The reduction of the delay variation as the gate size increases has been used to improve digital circuits delay robustness against process variations [46, 47].

Hint *Since the standard deviation of the inverter delay (σ_D) has a quadratic dependence on the gate size, sizing up the gate is an efficient way to reduce σ_D.*

4.6.2 Load Capacitance

Equation (4.8) shows that the delay sensitivities, and hence the standard deviation of the inverter delay, strongly depend on the load capacitance. The delay sensitivities increase linearly as the load capacitance increases, and as a consequence, the standard deviations increase [24, 44, 48]. The increases of the standard deviations with the load capacitance can be appreciated in Fig. 4.9. Nodes with high capacitance values are nodes with higher fan-out and/or long wire routing.

Hint *Care should be taken to those nodes with higher capacitance values to reduce their impact on the variance of the gate delay.*

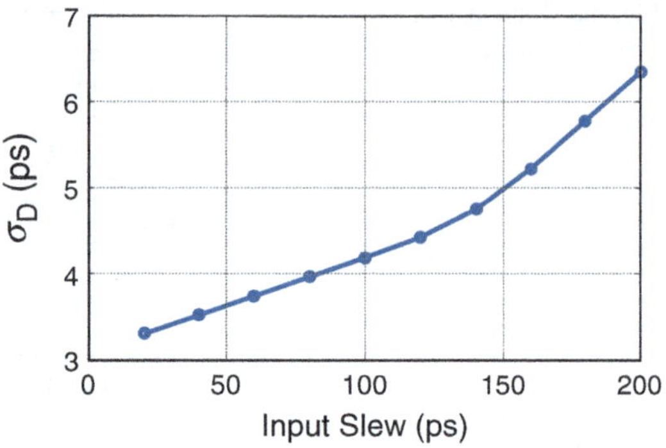

Fig. 4.10 Impact of input rise time on the standard deviation of the delay

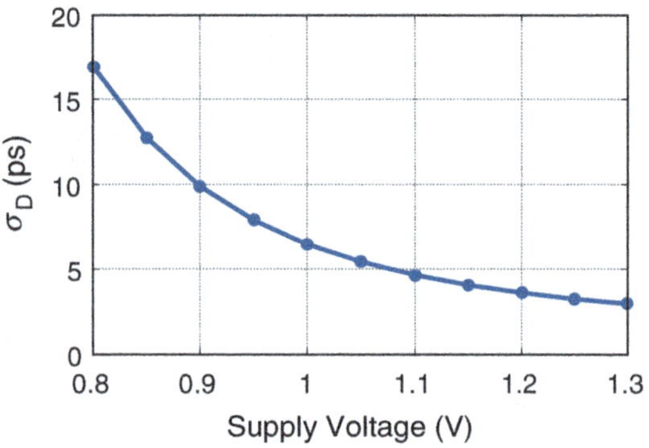

Fig. 4.11 Influence of power supply voltage

4.6.3 Input Slew Time

Figure 4.10 plots the standard deviation of the inverter delay versus the input rise time for a single inverter. The standard deviation of the inverter delay increases as the input rise time increases [44]. The increase of the standard deviation of the inverter delay can be even more significant for larger values of the input rise time. The standard deviation of the inverter delay increases because the nMOS network becomes more resistive for larger values of the input rise time.

Hint *Keep small where possible the values of the input transition times to reduce their impact on the variance of the gate delay.*

4.6.4 Power Supply Voltage

Equation (4.8) shows that the delay sensitivities, and hence the standard deviation of the inverter delay, strongly depend on the power supply voltage. The standard deviation of the delay increases as the power supply voltage is reduced [23, 26, 44]. This behavior can be appreciated in Fig. 4.11 where the standard deviation of the

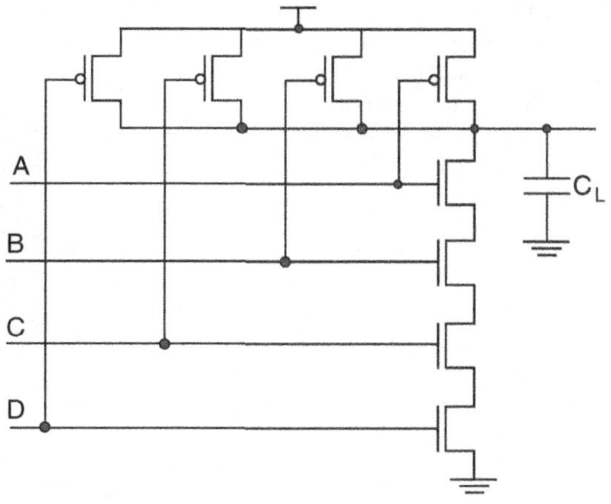

Fig. 4.12 Schematic circuit of a 4-Nand gate

inverter delay versus the power supply voltage is plotted. The impact of the power supply voltage on the standard deviation of the delay is more significant at lower values of the power supply than at higher values.

4.6.5 Gate Delay Dependence on Input Transition

The dependence of the behavior of the standard deviation of the delay of a 4-Nand gate (Fig. 4.12) on the input transitions is analyzed [44, 49]. The input transitions affect the strength of the switching network of the 4-Nand gate, which translates to the standard deviation of the delay of the gate.

Figure 4.13a shows the standard deviation of the delay of the 4-Nand gate (Fig. 4.12) for the high-to-low propagation delay. For a single input transition, the standard deviation of the delay depends on the position of the input transition on the stack of the nMOS switching network [49]. The standard deviation of the gate delay is smaller for an input transition closer to the output than for an input transition farther. Even more, the standard deviation of the delay is even larger for a simultaneous transition at the four inputs.

Figure 4.13b shows the standard deviation of the delay of the 4-Nand gate for the low-to-high propagation delay for single and simultaneous input transitions. The standard deviation of the delay for a single input transition is larger than that of simultaneous transition at the four inputs.

4.7 Summary

In this chapter, a mathematical formulation of the statistical delay of a gate was presented. The statistical performance of the inverter gate for correlated parameters and pure random variations was analyzed. Then, the analysis was extended to a

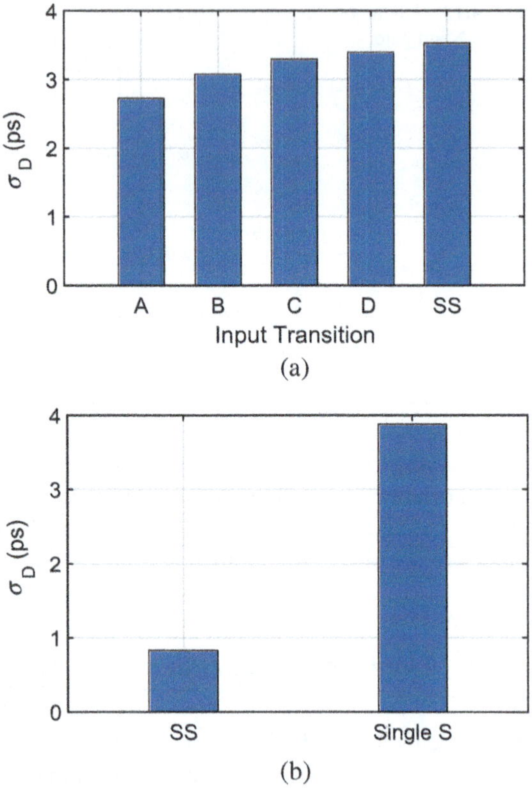

Fig. 4.13 Dependence of the standard deviation of the delay of a 4-Nand gate on the switching input transitions. (**a**) High-to-low propagation. (**b**) Low-to high propagation delay

Nand gate and generalized to other types of logic gates such as Nor, EX-OR, and AOI/AIO gates. Hand-by-hand examples to compute the delay variance (standard deviation) of logic gates were presented. Several key design issues affecting the gate delay were illustrated such as sizing a logic gate, load capacitance, input load capacitance, input slew time, power supply voltage, and gate delay dependence on input transition.

The gate delay variance is determined by (key concepts to remember):

- The delay sensitivities of each correlated parameter and each with pure random variations.
- The delay sensitivities are computed according to the input transitions of the gate.
- The standard deviations of the process parameters that are usually provided by the foundry.
- A straightforward way to modify the gate delay sensitivities, and thus the gate delay standard deviation, is by modifying the gate size.

References

1. C. Forzan, D. Pandini, Why we need statistical static timing analysis, in *25th International Conference on Computer Design*, Lake Tahoe, Oct 2007 (2007), pp. 91–96. https://doi.org/10.1109/ICCD.2007.4601885
2. S.R. Nassif, Modeling and analysis of manufacturing variations, in *IEEE Custom Integrated Circuits Conference* (2001), pp. 223–228. https://doi.org/10.1109/CICC.2001.929760
3. Y. Zhan, A.J. Strojwas, X. Li, L.T. Pileggi, D. Newmark, M. Sharma, Correlation-aware statistical timing analysis with non-Gaussian delay distributions, in *Proceedings 42nd Design Automation Conference*, Anaheim, June 2005 (2005), pp. 77–82. https://doi.org/10.1145/1065579.1065605
4. R. Goyal, S. Shrivastava, H. Parameswaran, P. Khurana, Improved first-order parameterized statistical timing analysis for handling slew and capacitance variation, in *20th International Conference on VLSI Design Held Jointly with 6th International Conference on Embedded Systems (VLSID'07)*, Bangalore, Jan 2007, pp. 278–282. https://doi.org/10.1109/VLSID.2007.92
5. S. Abbaspour, H. Fatemi, M. Pedram, VGTA: variation-aware gate timing analysis, in *2005 International Conference on Computer Design*, Oct 2005, pp. 351–356. https://doi.org/10.1109/ICCD.2005.115
6. A. Agarwal, D. Blaauw, V. Zolotov, S. Sundareswaran, M. Zhao, K. Gala, R. Panda, Path-based statistical timing analysis considering inter-and intra-die correlations, in *Proceedings of TAU* (2002), pp. 16–21
7. K. Kang, B.C. Paul, K. Roy, Statistical timing analysis using levelized covariance propagation, in *41st Design, Automation and Test in Europe Conference*, Mar 2005, vol. 2, pp. 764–769. https://doi.org/10.1109/DATE.2005.279
8. M. Orshansky, S. Nassif, D. Boning, *Design for Manufacturability and Statistical Design: A Constructive Approach* (Springer, Boston, 2008)
9. S.S. Sapatnekar, Variability and statistical design. IPSJ Trans. Syst. LSI Des. Methodol. **1**, 18–32 (2008)
10. A. Papoulis, S.U. Pillai, *Probability, Random Variables and Stochastic Processes*, 4th edn. (McGraw-Hill, Boston, 2002)
11. H. Mahmoodi, S. Mukhopadhyay, K. Roy, Estimation of delay variations due to random-dopant fluctuations in nanoscale CMOS circuits. IEEE J. Solid-State Circuits **40**(9), 1787–1796 (2005). https://doi.org/10.1109/JSSC.2005.852164
12. D. Blaauw, K. Chopra, A. Srivastava, L. Scheffer, Statistical timing analysis: from basic principles to state of the art. IEEE Trans. Comput. Aided Des. Integr. Circuits Syst. **27**(4), 589–607 (2008). https://doi.org/10.1109/TCAD.2007.907047
13. C. Forzan, D. Pandini, Statistical static timing analysis: a survey. Integr. VLSI J. **42**(3), 409–435 (2009). ISSN 0167-9260. Special Section on DCIS2006. https://doi.org/10.1016/j.vlsi.2008.10.002
14. T. Sakurai, A.R. Newton, Alpha-power law MOSFET model and its applications to CMOS inverter delay and other formulas. IEEE J. Solid-State Circuits **25**, 584–594 (1990). https://doi.org/10.1109/4.52187
15. SPICE2: a Computer Program to Simulate Semiconductor Devices, Memo ERL-M520, Department of Electrical Engineering and Computer Science, University of California at Berkeley (1975)
16. H. Chang, S.S. Sapatnekar, Statistical timing analysis under spatial correlations. IEEE Trans. Comput. Aided Des. Integr. Circuits Syst. **24**(9), 1467–1482 (2005). https://doi.org/10.1109/TCAD.2005.850834
17. D.C. Montgomery, G.C. Runger, *Applied Statistics and Probability for Engineers* (Wiley, Hoboken, 2006)

18. S. Bhardwaj, P. Ghanta, S. Vrudhula, A framework for statistical timing analysis using non-linear delay and slew models, in *2006 IEEE/ACM International Conference on Computer Aided Design*, San Jose, Nov 2006, pp. 225–230. https://doi.org/10.1109/ICCAD.2006.320140
19. L. Cheng, J. Xiong, L. He, Non-linear statistical static timing analysis for non-gaussian variation sources, in *2007 44th ACM/IEEE Design Automation Conference*, San Diego, June 2007, pp. 250–255
20. J. Singh, S.S. Sapatnekar, A scalable statistical static timing analyzer incorporating correlated non-gaussian and gaussian parameter variations. IEEE Trans. Comput. Aided Des. Integr. Circuits Syst. **27**(1), 160–173 (2008). https://doi.org/10.1109/TCAD.2007.907241
21. L. Cheng, J. Xiong, L. He, Non-gaussian statistical timing analysis using second-order polynomial fitting. IEEE Trans. Comput. Aided Des. Integr. Circuits Syst. **28**(1), 130–140 (2009). https://doi.org/10.1109/TCAD.2008.2009143
22. Z. Feng, P. Li, Y. Zhan, Fast second-order statistical static timing analysis using parameter dimension reduction, in *2007 44th ACM/IEEE Design Automation Conference*, San Diego, June 2007, pp. 244–249
23. M. Eisele, J. Berthold, D. Schmitt-Landsiedel, R. Mahnkopf, The impact of intra-die device parameter variations on path delays and on the design for yield of low voltage digital circuits. IEEE Trans. Very Large Scale Integr. Syst. **5**(4), 360–368 (1997). https://doi.org/10.1109/92.645062
24. B.P. Wong, A. Mittal, Y. Cao, G. Starr, *Nano-CMOS Circuit and Physical Design* (Wiley, Hoboken, 2005). https://doi.org/10.1002/0471653829.ch11
25. H. Onodera, Variability modeling and impact on design, in *IEEE International Electron Device Meeting*, Dec 2008, pp. 773–778. https://doi.org/10.1109/IEDM.2008.4796791
26. X. Zhang, X. Bai, Process variability-induced timing failures – a challenge in nanometer CMOS low-power design, in *Emerging Technologies and Circuits*, ed. by A. Amara, T. Ea, M. Belleville. Lecture Notes in Electrical Engineering, vol. 66 (Springer, Dordrecht, 2010). https://doi.org/10.1007/978-90-481-9379-012
27. J.K. Lorenz et al., Hierarchical simulation of process variations and their impact on circuits and systems: results. IEEE Trans. Electron Devices **58**(8), 2227–2234 (2011). https://doi.org/10.1109/TED.2011.2150226
28. C. Hawkins, J. Segura, P. Zarkesh-Ha, *CMOS Digital Integrated Circuits A First Course* (SciTech Publishing, Edison, New Jersey 2013)
29. H. Chang, S.S. Sapatnekar, Statistical timing analysis considering spatial correlations using a single pert-like traversal, in *ICCAD-2003 International Conference on Computer Aided Design*, Washington, DC, Nov 2003, pp. 621–625. https://doi.org/10.1109/ICCAD.2003.159746
30. C.S. Amin et al., Statistical static timing analysis: how simple can we get?, in *Proceedings 42nd Design Automation Conference*, New York, NY (Association for Computing Machinery, New York, 2005). pp. 652–657. https://doi.org/10.1145/1065579.1065751
31. H. Masuda, S. Ohkawa, A. Kurokawa, M. Aoki, Challenge: variability characterization and modeling for 65- to 90-nm processes, in *Proceedings of the IEEE 2005 Custom Integrated Circuits Conference*, Sep 2005, pp. 593–599. https://doi.org/10.1109/CICC.2005.1568738
32. W. Verhaegen, G. Van der Plas, G. Gielen, Automated test pattern generation for analog integrated circuits, in *Proceedings of the 15th IEEE VLSI Test Symposium*, May 1997
33. A. Srivastava, D. Sylvester, D. Blaauw, Statistical optimization of leakage power considering process variations using dual-Vth and sizing, in *Proceedings of the 41st Annual Design Automation Conference*, New York, NY (Association for Computing Machinery, New York, 2004), pp. 773–778. https://doi.org/10.1145/996566.996775
34. K.A. Gupta, D.K. Anvekar, V. Venkateswarlu, Comparative study and analysis of short channel effects for 180 nm and 45 nm transistors, in *Advances in Computing and Information Technology*, ed. by N. Meghanathan, D. Nagamalai, N. Chaki. Advances in Intelligent Systems and Computing, vol. 178 (Springer, Berlin, 2013). https://doi.org/10.1007/978-3-642-31600-569

35. K. Okada, K. Yamaoka, H. Onodera, Statistical gate-delay modeling with intra-gate variability, in *ICCAD-2003 International Conference on Computer Aided Design*, Nov 2003, pp. 908–913. https://doi.org/10.1109/ICCAD.2003.159782
36. E. Acar, H. Mau, A. Heinig, B. Li, U. Schlichtmann, Examination of process parameter variations, in *Process Variations and Probabilistic Integrated Circuit Design*, ed. by M. Dietrich, J. Haase (Springer, New York, 2012). https://doi.org/10.1007/978-1-4419-6621-6_3
37. K. Shinkai, M. Hashimoto, T. Onoye, A gate-delay model focusing on current fluctuation over wide range of process-voltage-temperature variations. Integr. VLSI J. **46**, 345–358 (2013). https://doi.org/10.1016/j.vlsi.2013.01.003
38. V. Litovsky, M. Zwolinsky, *VLSI Circuit Simulation and Optimization* (Springer Science and Business Media, Boston, 1996)
39. L. Guerra e Silva, Z. Zhu, J.R. Phillips, L.M. Silveira, Library compatible variational delay computation, in *VLSI-SoC Research Trends in VLSI and Systems on Chip*, ed. by G. De Micheli, S. Mir, R. Reis (Springer, New York, 2008). https://doi.org/10.1007/978-0-387-74909-9_10
40. J. Xiong, V. Zolotov, L. He, Robust extraction of spatial correlation. IEEE Trans. Comput. Aided Des. Integr. Circuits Syst. **26**(4), 619–631 (2007). https://doi.org/10.1109/TCAD.2006.884403
41. J. Xiong, V. Zolotov, C. Visweswariah, Efficient modeling of spatial correlations for parameterized statistical static timing analysis, in *2009 IEEE 8th International Conference on ASIC*, Changsha, Hunan, Oct 2009, pp. 722–725 (2009). https://doi.org/10.1109/ASICON.2009.5351307
42. J.P. Uyemura, *CMOS Logic Circuit Design* (Kluwer, Boston, 2002)
43. D. Hodges, H. Jackson, R. Saleh, *Analysis and Design of Digital Integrated Circuits*, 3rd edn. (McGraw-Hill, New York, 2004)
44. L. Yu, S. Saxena, C. Hess, I.A.M. Elfadel, D. Antoniadis, D. Boning, Statistical library characterization using belief propagation across multiple technology nodes, in *2015 Design, Automation Test in Europe Conference Exhibition (DATE)*, Mar 2015, pp. 1383–1388
45. M. Alioto, G. Palumbo, M. Pennisi, Understanding the effect of process variations on the delay of static and domino logic. IEEE Trans. Very Large Scale Integr. Syst. **18**(5), 697–710 (2010). https://doi.org/10.1109/TVLSI.2009.2015455
46. A. Agarwal, K. Chopra, D. Blaauw, Statistical timing based optimization using gate sizing, in *Design, Automation and Test in Europe*, Mar 2005, vol. 1, pp. 400–405. https://doi.org/10.1109/DATE.2005.281
47. S.H. Choi, B.C. Paul, K. Roy, Novel sizing algorithm for yield improvement under process variation in nanometer technology, in *Proceedings of the 41st annual Design Automation Conference 2004*, July 2004, New York, pp. 454–459
48. J.B. Sulistyo, D.S. Ha, A new characterization method for delay and power dissipation of standard library cells. VLSI Des. **15**(3), 667–678 (2002). https://doi.org/10.1080/1065514021000012273
49. D.N. da Silva, A.I. Reis, R.P. Ribas, CMOS logic gate performance variability related to transistor network arrangements. Microelectron. Reliab. **49**(9), 977–981 (2009). https://doi.org/10.1016/j.microrel.2009.07.023

Chapter 5
Path Delay Under Process Variations

Abstract The previous chapter described the statistical delay under process variations of a digital gate. This chapter studies the statistical performance of a logic path composed of several gates under process variations. It is shown how to compute the delay variance (standard deviation) of a logic path. This is illustrated in detail for a two-inverter chain, and then this analysis is extended to a general logic path. Pure random and correlated variations are considered. The behavior of a logic path under process variations is analyzed. Several key design issues affecting the path delay are illustrated such as the impact of spatial correlation, the influence of relative delay sensitivities between gates, sizing a logic path, and logic depth. Based on the previous key design issues, designers can take actions to improve their designs and fulfill timing specifications efficiently. In the last issue addressed in this chapter, the main advantages of statistical-based design over corner-based design are analyzed. Overheads due to pessimistic corner design are increasing and becoming non-tolerable in nanometer digital circuits.

5.1 Computing Statistical Delay of a Logic Path

The statistical delay of a two-inverter chain is computed based on the sensitivity delay model, and then the results are extended to the statistical delay of a general logic path. An example step by step to compute the statistical delay of a logic path is presented.

5.1.1 Computing Statistical Delay of a Two-Inverter Chain

Figure 5.1 shows the schematic circuit of a two-inverter chain illustrating possible logic path trajectories depending on the input transition. A low- to high-input transition creates a logic path through the nMOS network of inverter *1* and the pMOS network of inverter *2*. On the other hand, a high- to low-input transition creates a logic path through the pMOS network of inverter *1* and the nMOS network of inverter *2*.

V. Champac, J. G. Gervacio, *Timing Performance of Nanometer Digital Circuits Under Process Variations*, Frontiers in Electronic Testing 39,
https://doi.org/10.1007/978-3-319-75465-9_5

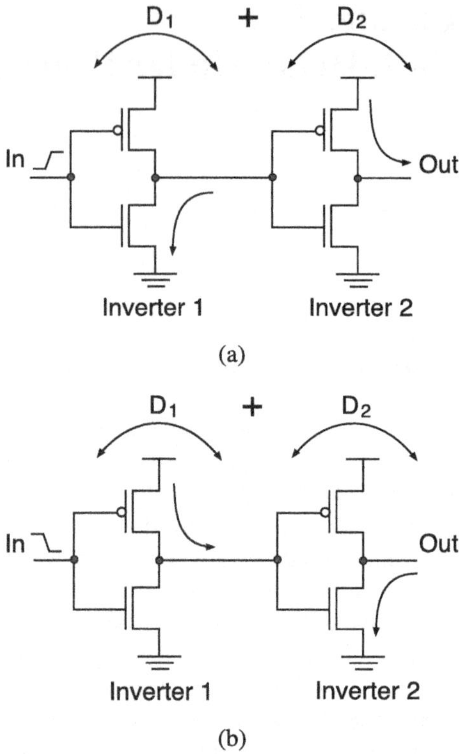

Fig. 5.1 Logic paths for an inverter gate. (**a**) Logic path for a low- to high-input transition. (**b**) Logic path for a high- to low-input transition

Normal distributions are assumed for the delay of the inverters, D_1 ($\sim N(\mu_{D_1}, \sigma^2_{D_1})$) and D_2 ($\sim N(\mu_{D_2}, \sigma^2_{D_2})$). For the low- to high-input transition (Fig. 5.1a), D_1 corresponds to the gate delay introduced by the nMOS network of inverter *1*, and D_2 corresponds to the gate delay introduced by the pMOS network of inverter *2*. For the high- to low-input transition, the roles of the networks are reversed.

The statistical delay of an inverter can be obtained using Eqs. (4.2) and (4.5), which are based on the gate linear model [1–6]. The mean and variance of the delay for inverter 1 are expressed as

$$\mu_{D_1} = D_{1,0} \tag{5.1}$$

$$\sigma^2_{D_1} = \sum_{m=1}^{k} S^2_{D_{1,m}} \sigma^2_{D_{1,m}} \tag{5.2}$$

Similarly, the statistical delay of inverter 2 is expressed as

$$\mu_{D_2} = D_{2,0} \tag{5.3}$$

$$\sigma_{D_2}^2 = \sum_{m=1}^{k} S_{D_{2,m}}^2 \sigma_{D_{2,m}}^2 \tag{5.4}$$

The following assumptions are made for estimating the path delay:

Assumption 1 It is assumed that nonzero spatial correlations exist for the same type of parameter, while correlations do not exist among different types of parameters [7].

Assumption 2 The delay distribution of a gate due to the delay distribution of the signal arrival time is not considered [8].

Let us remember how to sum random variables. The mean delay value of the sum of n random variables is obtained by summing up the mean values of the normal distributions of the variables, and the variance of the sum is obtained by summing their variances with all their covariances [9]. The obtained output delay is also a normal distribution [9, 10]. Let us compute the mean and standard deviation of a logic path composed of a two-inverter chain (Fig. 5.1). The mean delay at the output of the chain is computed as the accumulation of all the nominal gate delays in the path. The mean delay at the output of the path (μ_{Dp}) is given by

$$\mu_{Dp} = \mu_{D_1} + \mu_{D_2} = D_{1,0} + D_{2,0} \tag{5.5}$$

The mean delay (μ_D) of each gate is evaluated using SPICE electrical simulator to have better accuracy.

The variance at the output of the chain is obtained by the sum of two correlated random variables as given in Eq. (2.25) [9–11]. The variance of the delay at the output of the path (σ_{Dp}^2) is given by

$$\sigma_{Dp}^2 = \sigma_{D_1}^2 + \sigma_{D_2}^2 + 2\mathrm{Cov}(D_1, D_2) \tag{5.6}$$

where the first two terms are the variances of each inverter and $\mathrm{Cov}(D_1, D_2)$ is the covariance between their delay distributions.

The variance of the delay of each inverter ($\sigma_{D_1}^2, \sigma_{D_2}^2$) in Eq. (5.6) is obtained with Eqs. (5.2) and (5.4) using the delay sensitivity model (Sect. 4.3.2 in Chap. 4).

The covariance between delays in Eq. (5.6) is rewritten in terms of the delay sensitivities and the correlation between parameters. Two steps are performed.

Step 1 The covariance between the delay distributions of the two inverters is expressed in terms of the delay sensitivities to each parameter and the covariances between the distributions of the parameters [12] such as given in Eq. (A.9) of Appendix A. Only parameters presenting spatial correlation (W, L, and T_{ox}) are considered, and V_{th} is not considered as this parameter is assumed to present pure random variations. Equation (5.7) is obtained assuming a linear combination of the random variables representing the process parameters.

$$\begin{aligned}\mathrm{Cov}(D_1, D_2) = {} & S_{D_1,W} S_{D_2,W} \cdot \mathrm{Cov}(X_{W,D_1}, X_{W,D_2}) \\ & + S_{D_1,L} \cdot S_{D_2,L} \cdot \mathrm{Cov}(X_{L,D_1}, X_{L,D_2}) \\ & + S_{D_1,T_{ox}} \cdot S_{D_2,T_{ox}} \cdot \mathrm{Cov}(X_{T_{ox},D_1}, X_{T_{ox},D_2})\end{aligned} \tag{5.7}$$

where $S_{D_1,W}$ and $S_{D_2,W}$ are the delay sensitivities to small changes in W of inverters *1* and 2, respectively. $S_{D_1,L}$, $S_{D_2,L}$, $S_{D_1,T_{ox}}$, and $S_{D_2,T_{ox}}$ are similarly defined. $\mathrm{Cov}(X_{W,D_1}, X_{W,D_2})$ is the covariance between the distributions of the parameter W for inverters *1* and *2*. $\mathrm{Cov}(X_{L,D_1}, X_{L,D_2})$ and $\mathrm{Cov}(X_{T_{ox},D_1}, X_{T_{ox},D_2})$ are similarly defined.

Step 2 The *covariances* between the distributions of two parameters in Eq. (5.7) are translated to *correlations* between the parameters according to Eq. (2.21) [10, 11]. For a parameter P_m, this is given by

$$\mathrm{Cov}(X_{P_m,D_1}, X_{P_m,D_2}) = \rho_{Pm_{D_1,D_2}} \cdot \sigma_{P_m,D_1} \cdot \sigma_{P_m,D_2} = \rho_{Pm_{D_1,D_2}} \cdot \sigma_{P_m} \cdot \sigma_{P_m} \tag{5.8}$$

where $\rho_{Pm_{D_1,D_2}}$ is the correlation of the distributions of a parameter P_m for the devices located at inverters *1* and *2*. σ_{P_m,D_1} is the standard deviation of parameter m at the inverter *1* and σ_{P_m,D_2} is the standard deviation of parameter P_m at the inverter *2*. It should be noted that $\sigma_{P_m,D_1} = \sigma_{P_m,D_2}$ because they refer to the same process parameter.

Replacing Eq. (5.8) in Eq. (5.7) for each process parameter, the covariance between the delay distributions of the two inverters is expressed as

$$\begin{aligned}\mathrm{Cov}(D_1, D_2) = {} & S_{D_1,W} S_{D_2,W} \cdot \rho_{W_{D_1,D_2}} \cdot \sigma_W \cdot \sigma_W \\ & + S_{D_1,L} \cdot S_{D_2,L} \cdot \rho_{L_{D_1,D_2}} \cdot \sigma_L \cdot \sigma_L \\ & + S_{D_1,T_{ox}} \cdot S_{D_2,T_{ox}} \cdot \rho_{T_{ox}D_1,D_2} \cdot \sigma_{T_{ox}} \cdot \sigma_{T_{ox}}\end{aligned} \tag{5.9}$$

Finally, replacing Eq. (5.9) in Eq. (5.6), the expression to compute the variance of the delay at the output of a two-inverter chain is given by

$$\sigma_{Dp}^2 = \sigma_{D_1}^2 + \sigma_{D_2}^2 + 2 \sum_{m=1}^{k} S_{D_1,P_m} \cdot S_{D2,P_m} \cdot \rho_{Pm_{D_1,D_2}} \cdot \sigma_{P_m} \cdot \sigma_{P_m} \tag{5.10}$$

where the third term considers all the parameters presenting spatial correlation.

5.1.2 Computing Statistical Delay of a General Logic Path

The mean delay at the output of a general logic path composed of different type logic gates is obtained by Amin et al. [13],

$$\mu_{D_P} = \sum_{i=1}^{N} \mu_{D_i} \tag{5.11}$$

and an expression to compute the delay variance at the output of a general logic path composed of different types of logic gates is obtained by extending Eq. (5.6) [13]

$$\sigma_{D_P}^2 = \sum_{i=1}^{N} \sigma_{D_i}^2 + 2\sum_{i=1}^{N} \sum_{j=i+1}^{N} \text{Cov}(D_i, D_j) \tag{5.12}$$

with the covariances between the delay distributions of gates i and j expressed by

$$\text{Cov}(D_i, D_j) = \sum_{m=1}^{k} S_{Di,P_m} \cdot S_{Dj,P_m} \cdot \rho_{Pm_{Di,Dj}} \cdot \sigma_{P_m} \cdot \sigma_{P_m} \tag{5.13}$$

The first term in Eq. (5.12) considers the delay variances of all the N gates in the logic path, and the second term considers the covariance contributions due to the parameter spatial correlation between gate locations. The parameters in Eq. (5.13) have the same meaning than in Eq. (5.8).

The gate delay sensitivities are computed for the switching network. Let us consider a logic path activated for a single transition at the primary input with logic conditions at the inputs set as shown in Fig. 5.2. As a consequence of the input transition, the signals propagate from the input to the output of the path. The delay of each gate, modeled as normal distributions, corresponds to the activated switching network. For instance, the activated networks for the input transition shown in Fig. 5.2 are (1) nMOS network of the 2-Nand gate, (2) pMOS network of the inverter gate, (3) nMOS network of the 3-Nand gate, and (4) pMOS network of the 2-Nor gate. The delay of each gate is obtained following the methodology described in the previous chapter (Chap. 4). The delay at the output of the logic path is obtained by statistically summing up the gate delays due to the activated networks using Eqs. (5.12) and (5.13).

At the circuit level, the variance of the delay at the output of a logic path is determined by (Eq. (5.12)*):*

- The variance of each gate in the logic path.
- The covariances of the delay distributions between pairs of gates at the logic path.

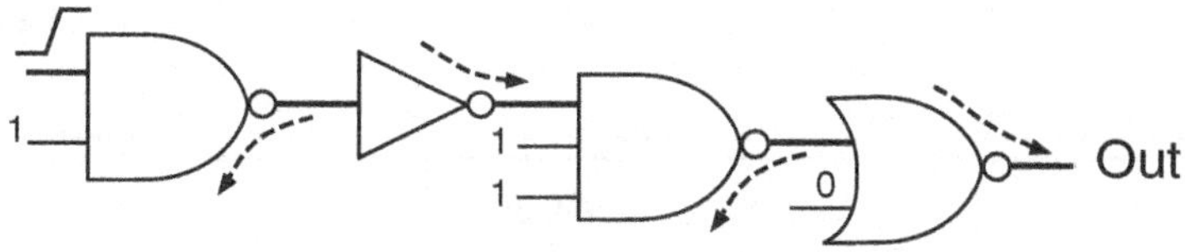

Fig. 5.2 Logic path

The covariance of the delay distributions between a pair of gates depends on (Eq. (5.13)*):*

- The delay sensitivities to small changes in parameters. Contributions of those parameters that suffer spatial correlation are considered. SPICE electrical simulator can be used for accurate computing of the delay sensitivities (Eq. (4.10)). First-order analytical expressions (Eqs. (4.8a)–(4.8d)) can also be used to compute delay sensitivities but with loss of accuracy.
- The spatial correlation of the parameters due to the physical distance that separates the gates (Eq. (3.13)). Layout information is required.

5.1.3 Example of Computing the Delay Variance of a Path

Figure 5.3 shows a logic path that is used to illustrate the procedure to compute the path delay. The gates are designed symmetrically with minimum channel length. The inverters have $W_N = 120\,\text{nm}$ and $W_P = 240\,\text{nm}$, the Nor gate has $W_N = 120\,\text{nm}$ and $W_P = 480\,\text{nm}$, and the Nand gate has $W_N = 240\,\text{nm}$ and $W_P = 240\,\text{nm}$. The load capacitance of each gate is indicated in Fig. 5.3. Standard deviations of the process parameters W, L, T_{ox}, and V_{th} are the same than those given in Table 4.1. The transistor threshold voltages accounting the effect of RDF are calculated using Eq. (3.2).

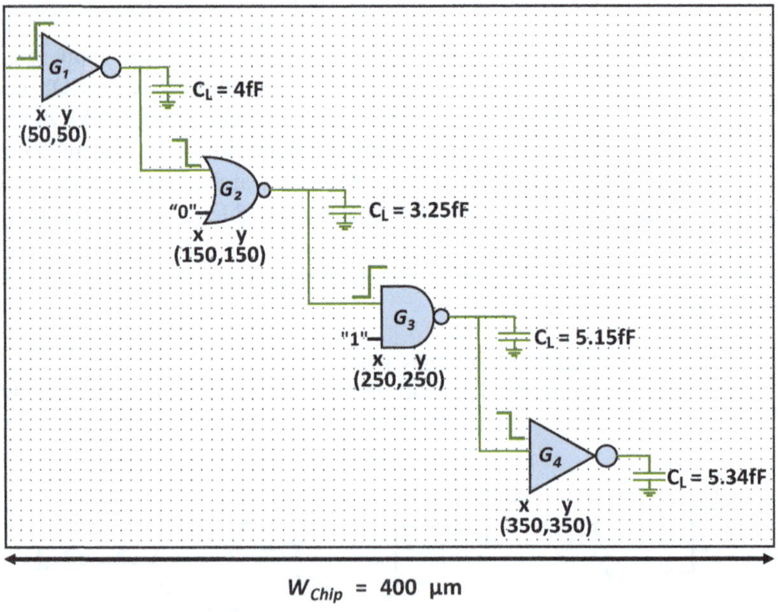

Fig. 5.3 Circuit used to illustrate path delay computing, $CD_{\text{WID}} = W_{\text{chip}}/2$

The delay variance of a logic path is computed using Eqs. (5.12) and (5.13). These equations require to compute the delay variance of each gate and the covariances between gates.

Computing Gate Delay Variances

The delay variance of each gate is calculated using Eq. (4.21) repeated next for convenience,

$$\sigma_{D_{\text{tot}}}^2 = S_{D,W}^2\sigma_W^2 + S_{D,L}^2\sigma_L^2 + S_{D,T_{\text{ox}}}^2\sigma_{T_{\text{ox}}}^2 + \sum_{i=1}^{p} S_{D,V_{\text{th}_i}}\sigma_{V_{\text{th}_i}} \quad (5.14)$$

The delay sensitivities are estimated using SPICE electrical simulator and are used to calculate both the variance of each gate (Eq. (5.14)) and the covariances between gates (Eq. (5.13)). For the serially connected networks in the 2-Nor and 2-Nand gates, the delay sensitivities of the correlated parameters are computed with the compact delay model. For inverters G_1 and G_4, the delay sensitivities are obtained for the high- to low-output transition and low- to high-output transition, respectively. For gates G_2 and G_3, the delay sensitivities are obtained for the low- to high- and high- to low-output transitions, respectively, with the input conditions given in Fig. 5.3. For the parameter (V_{th}) having pure random variations, the delay sensitivities are computed separately for each one of the activated transistors according to the input conditions given in Fig. 5.3. The delay sensitivities of the gates in the logic path (Fig. 5.3) estimated using SPICE electrical simulator are shown in Table 5.1.

By using the standard deviations of technological parameters (Table 4.1) and the gate delay sensitivities given in Table 5.1, the delay variances of each gate of the logic path are computed using Eq. (5.14) (Table 5.2).

Table 5.1 Gate delay sensitivities for the logic path shown in Fig. 5.3

Gate	Input	$S_{D,W}(s/m)$	$S_{D,L}(s/m)$	$S_{D,T_{\text{ox}}}(s/m)$	$S_{D,V_{\text{th}},M1}\ (s/V)$	$S_{D,V_{\text{th}},M2}\ (s/V)$
G_1	$0 \to 1$	−2.9437E−04	3.0178E−04	8.6878E−03	1.7853E−10	–
G_2	$1 \to 0$	−0.6434E−04	5.1115E−04	9.8315E−03	1.4936E−10	9.4783E−11
G_3	$0 \to 1$	−1.9201E−04	5.9718E−04	13.235E−03	1.3324E−10	1.8225E−10
G_4	$1 \to 0$	−1.4219E−04	3.9516E−04	9.9415E−03	2.0236E−10	–

Table 5.2 Variance of the delay of each gate in the example

Gate	σ_D^2 (ps^2)
G_1	47.937
G_2	13.941
G_3	41.341
G_4	29.821

Computing Covariances Between Delay Distributions

The covariances between delay distributions of a pair of gates are computed using Eq. (5.13). The gate delay sensitivities are those given in Table 5.1, and the standard deviations of technological parameters are given in Table 4.1. The spatial correlation of a parameter P_m for two gates located at the positions i and j is estimated with Eq. (3.13) repeated below (Eq. (5.15)) for convenience. The meaning of each one of the terms of Eq. (5.15) is explained in Chap. 3. This equation requires knowing K_{D2D}, K_{WID}, $\mathrm{CD}_{\mathrm{WID}}$, CD_{D2D}, and d_{ij}. It is assumed that the parameters W, L, and T_{ox} present inter-die and intra-die correlated variations without intra-die pure random variation component. 20% ($K_{D2D} = 0.2$) for inter-die correlated variations and 80% ($K_{\mathrm{WID}} = 0.8$) for intra-die correlated variations have been considered. The die width (W_{chip}) given in Fig. 5.3 is used to compute $\mathrm{CD}_{\mathrm{WID}}$ ($\mathrm{CD}_{\mathrm{WID}} = W_{\mathrm{chip}}/2$). The distance between a pair of gates (d_{ij}) is calculated using the gate coordinates given in Fig. 5.3.

$$\rho_{Pm_{ij}} = K_{D2D} + K_{\mathrm{WID}} \cdot \exp\left(-\frac{d_{ij}}{\mathrm{CD}_{\mathrm{WID}}}\right) \tag{5.15}$$

Matrix (5.16) shows the calculated spatial correlations using (5.15) between a pair of gates of the logic path shown in Fig. 5.3. Each position in the matrix corresponds to the spatial correlation of the parameter between gates G_i and G_j (Fig. 5.3).

$$\rho_{m_{ij}} = \begin{array}{c} \begin{matrix} G_1 & G_2 & G_3 & G_4 \end{matrix} \\ \begin{pmatrix} 1.0000 & 0.5945 & 0.3945 & 0.2959 \\ 0.5945 & 1.0000 & 0.5945 & 0.3945 \\ 0.3945 & 0.5945 & 1.0000 & 0.5945 \\ 0.2959 & 0.3945 & 0.5945 & 1.0000 \end{pmatrix} \end{array} \begin{matrix} G_1 \\ G_2 \\ G_3 \\ G_4 \end{matrix} \tag{5.16}$$

By replacing the standard deviations of technological parameters (Table 4.1), the delay sensitivities (Table 5.1), and the spatial correlation Matrix (5.16) in Eq. (5.13), the covariances between each pair of gates of the logic path are computed. The obtained covariances are given in Matrix (5.17). It should be noted that Matrix (5.17) also includes the variances of each gate (diagonal of the matrix).

$$\mathrm{Cov}(D_i, D_j) = 1e^{-24} \cdot \begin{array}{c} \begin{matrix} D_1 & D_2 & D_3 & D_4 \end{matrix} \\ \begin{pmatrix} 47.9374 & 1.9618 & 2.0849 & 1.1314 \\ 1.9618 & 13.9421 & 3.0113 & 1.4116 \\ 2.0849 & 3.0113 & 41.3415 & 2.9730 \\ 1.1314 & 1.4116 & 2.9730 & 29.8215 \end{pmatrix} \end{array} \begin{matrix} D_1 \\ D_2 \\ D_3 \\ D_4 \end{matrix} \tag{5.17}$$

Computing Path Delay Variance

By replacing the delay variances of each gate (Table 5.2) and the delay covariances between gates (Matrix (5.17)) in Eq. (5.12), the delay variance of the logic path illustrated in Fig. 5.3 is computed. The standard deviation of the delay is obtained by the square root of the variance. The obtained standard deviation of the delay of the logic path with the delay sensitivity model is 12.57 ps. The standard deviation of the delay with a SPICE Monte Carlo simulation is 12.18 ps. Spatial correlations between logic gates were taken into account using the simulation methodology shown in Appendix C. The error between the computation using Eq. (5.12) and SPICE is 3.2%.

5.2 Circuit Issues Impacting Path Delay Under Process Variations

In this section, the behavior of a logic path under process variations is analyzed.

5.2.1 Impact of Spatial Correlation

The impact of the spatial correlation on path delay variance is discussed more in depth. Let us consider a two-inverter chain with load capacitances C_{L1} and C_{L2} (Fig. 5.4). The delay variance of the two-inverter chain is computed with Eq. (5.6) repeated below for convenience. The path delay variance depends on the contributions of the variances of each inverter and also on the covariances (due to spatial correlation) between the delays of the two gates. At the circuit level, the variance of each inverter is determined by their delay sensitivities, and the covariances between gate delays are determined by both the delay sensitivities and the separation distance between the gates (parameter ρ in Eq. (5.9)). It must be reminded that the delay sensitivities of the gates depend on their drive strengths and load capacitances (Eq. (4.8)). The delay sensitivities increase as the load capacitance of the gate increases. The load capacitances in the circuit of Fig. 5.4 are used to model the delay sensitivities of each inverter ($S_D = f(C_L)$).

$$\sigma_{Dp}^2 = \sigma_{D_1}^2 + \sigma_{D_2}^2 + 2\text{Cov}(D_1, D_2) \tag{5.18}$$

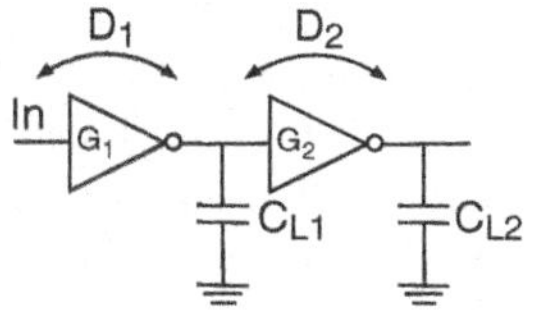

Fig. 5.4 A two-inverter chain

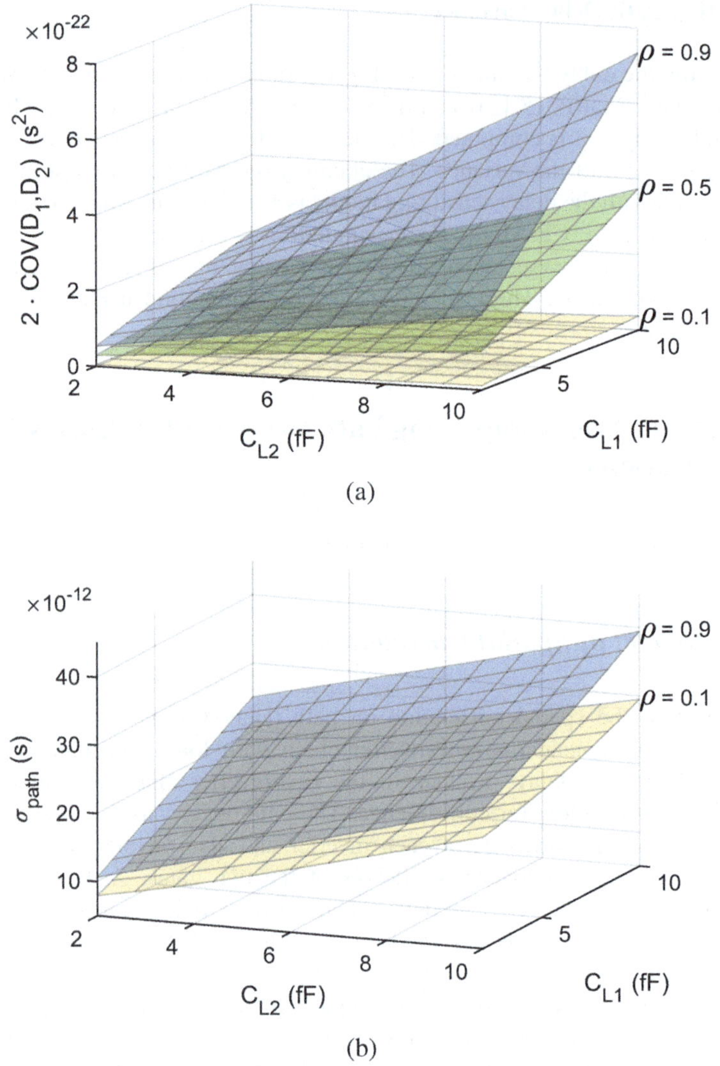

Fig. 5.5 Impact of spatial correlation on path delay, $S_{D_1} = f(C_{L1})$ and $S_{D_2} = f(C_{L2})$. (**a**) Covariance dependence on delay sensitivities. (**b**) Covariance impact on path delay

Figure 5.5a illustrates the variance contribution due to the spatial correlation (term $2\text{Cov}(D_1, D_2$ in Eq. (5.18)) to the path delay as a function of the load capacitances. For a given distance separation between the inverters modeled by ρ, the contribution of the covariance increases as the delay sensitivities (C_{L1}, C_{L2}) increase. The amount of the contribution of the spatial correlation is significantly larger for gates located very close ($\rho = 0.9$) than for gates located farther ($\rho = 0.1$).

Figure 5.5b plots the standard deviation of the path delay as a function of the load capacitances for inverters located very close ($\rho = 0.9$) and far away ($\rho = 0.1$). This figure illustrates the relative impact of the contribution of the spatial correlation in the overall path delay. An increase of the delay sensitivities of the inverters increases not only the contribution of the delay covariance term but also the contribution of the variance of each inverter (Eq. (5.18)). For instance, at $C_{L1} = 10fF$ and $C_{L2} = 10fF$, the standard deviation of the path delay with inverters located very close is around 1.8 times that with inverters located far away. And, at $C_{L1} = 2fF$ and $C_{L2} = 10fF$, the standard deviation of the path delay with inverters located very close is around 1.4 times that with inverters located far away. These observations in Fig. 5.5b give a further indication of the important role that the spatial correlation plays in the overall path delay variance. Because of this, the designer should account the spatial variance contribution for proper computing of the path delay.

5.2.2 *Relative Delay Sensitivities*

Previously, the significant impact of the delay sensitivities on path delay was shown. Let us see how this behavior can be used for improving path delay performance. Figure 5.6 shows a chain of two inverters located spatially close and also a chain of two inverters located spatially far. Figure 5.7 shows a bar graph of the standard

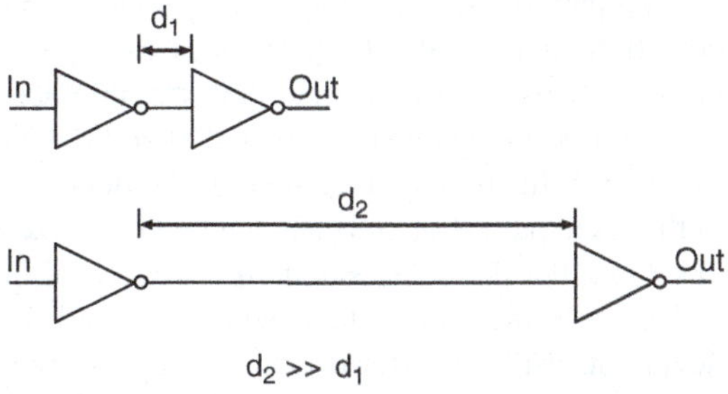

Fig. 5.6 Two-inverter chains with different spatial correlation. d_1 for a short distance separation ($\rho = 1$) and d_2 for a long distance separation ($\rho = 0$)

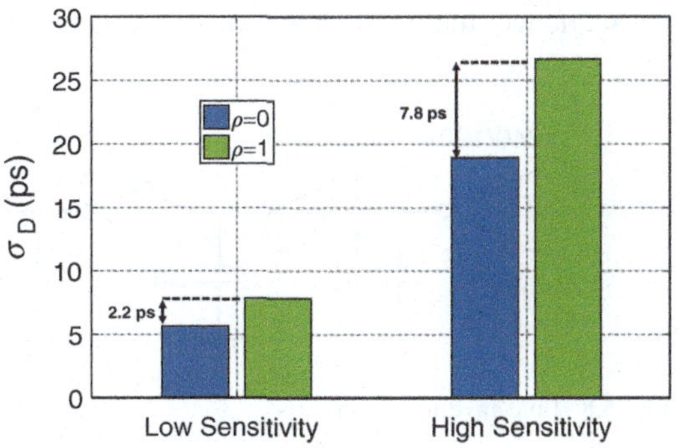

Fig. 5.7 Impact of delay sensitivities on path delay

deviation of the delay at the output of the two-inverter chain when both inverters have low delay sensitivities and also when they have high delay sensitivities. $\rho = 1$ is assumed for the two inverters located very close and $\rho = 0$ for the two inverters located far away. It is observed in Fig. 5.7 that the standard deviation of the delay at the output of the two-inverter chain increases when going from $\rho = 0$ to $\rho = 1$ for low and high delay sensitivities of the gates. Even more, the increase of the standard deviation of the delay when going from $\rho = 0$ to $\rho = 1$ is more significant when both inverters in the chain have high delay sensitivity than when they have low delay sensitivity.

Hint *Be aware of a pair of gates being located physically close and having high delay sensitivities. These conditions increase the variance contributions due to spatial correlation.*

5.2.3 Sizing a Logic Path

The impact of sizing up a gate on path delay is analyzed.

5.2.3.1 Behavior of Gate Sizing in a Logic Path

A chain of five inverters (Fig. 5.8) is used to illustrate the impact of sizing up a gate on path delay behavior. In the previous chapter, it was shown that the standard deviation of the gate delay reduces as a gate is sized up (Fig. 4.9). However, a gate is not isolated in a logic path but interacts with other gates. Sizing up a gate mainly modifies its own delay and also the delay of its driven and preceding gates. Figure 5.9 illustrates the change of the delay standard deviation of each gate ($\Delta\sigma_i$) in the logic path (Fig. 5.8) when gate G_3 is sized up by a certain amount (ΔK). It is observed that the delay standard deviation of the inverter G_3 reduces significantly, and it is the most important reduction of all the gates. Besides, the delay standard deviation of the inverter driven by gate G_3 (gate G_4) also reduces. This reduction is because the current delivered by gate G_3 increases as it is sized up, and as a consequence, the input slew time of G_4 is faster (Fig. 4.10).

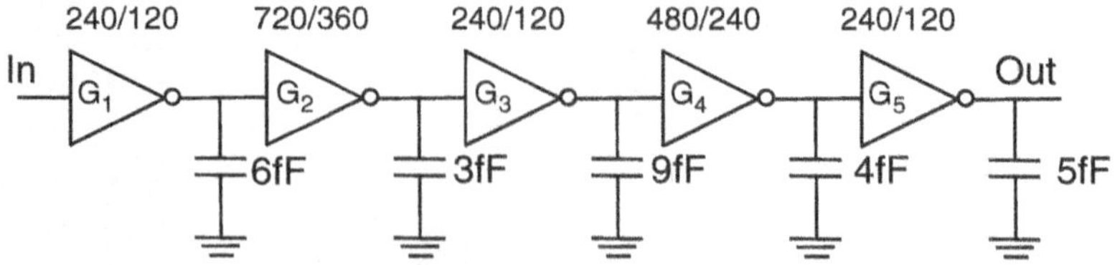

Fig. 5.8 Five-inverter chain

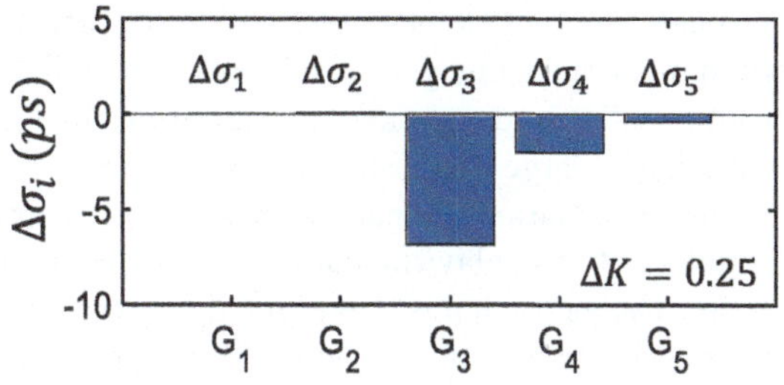

Fig. 5.9 Impact of sizing up a gate (G_3) on the standard delay deviation of each gate in a logic path

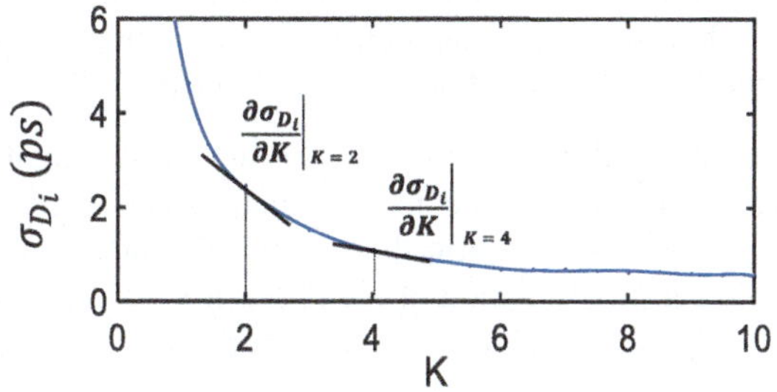

Fig. 5.10 Impact of sizing up a gate on its delay standard deviation

Even more, the reduction occurs in spite of the capacitance added to the input of G_4 due to sizing up gate G_3. Sizing up gate G_3 also increases the delay standard deviation of its preceding gate (gate G_2) and decreases the delay standard deviation of gate G_5, but these effects are less important than those observed in gates G_3 and G_4. Thus, the overall delay standard deviation of the path is determined not only by the reduction of delay standard deviation of the size up gate but also by the changes of the delay standard deviation on other gates in the path (Fig. 5.9).

5.2.3.2 Gate Sizing in a Logic Path

A designer is interested in identifying those gates of a logic path that when sized up have more reduction in the delay standard deviation of the path. Some metrics to select the best gates to size up have been proposed in the literature [14–16]. These metrics are usually used for design optimization of the overall circuit delay using CAD tools. A simple approach for guiding the designer for improving delay of specific critical paths is discussed next. A metric that evaluates how much reduces the delay standard deviation of gate when sized up for a certain amount is

$$\frac{\partial \sigma_{D_i}}{\partial K_i} \tag{5.19}$$

Figure 5.10 plots the delay standard deviation of an inverter with a fixed load capacitance as a function of its size. The delay standard deviation decreases as the size of the gate increases. It is observed that the ratio of the gate delay standard deviation change to the gate size increment decreases at higher values of K. For instance, the $\partial\sigma/\partial K$ at $K = 2$ is significantly larger than at $K = 4$. In other words, for a moderate same gate size increment, an inverter with size $K = 2$ reduces more significantly its delay standard deviation than an inverter with size $K = 4$. For a sufficiently large gate, an increase of the gate size barely reduces the gate delay standard deviation. A gate with a higher metric score (Eq. (5.19)) reduces more significantly its delay standard deviation in comparison to a gate with a lower metric score. The metric tends to present higher values for smaller gates. On the other hand, the metric tends to present lower values for a large gate, and thus, sizing up the gate does not provide real benefits of reduction of the delay standard deviation.

In order to use the metric (5.19) to select the best gates of a logic path to size up, the following simplifications are made:

1. The path is composed of the same type of gate.
2. Sizing up a gate has a much larger impact on the change of its own delay standard deviation than that in the other gates in the logic path. Note that this is a simplification of the observations in Fig. 5.9.
3. Spatial correlation is not considered.

Logic cells composing a path may have different driving capabilities and different load capacitances. Because of this, the score values obtained by the metric (5.19) may present different values for the gates in the logic path.

For the same size increment at the gates in the logic path, those gates with higher score provide a larger reduction of the gate delay standard deviation (Fig. 5.10), and hence, it can also be assumed in a first approach that they provide the higher reduction benefit in the path delay standard deviation. Figure 5.11 shows the percentage of optimization of the standard delay deviation of the logic path (%Opt σ_p) when sizing up a single gate at a time by the same amount (ΔK). It is observed that the degree of optimization is different depending on which gate is sized up. The gate providing the highest benefit of optimization of the standard delay deviation of the path corresponds to the gate with the highest score value of Eq. (5.19). The opposite is true for the gate providing the lowest benefit of optimization of the standard delay deviation of the path.

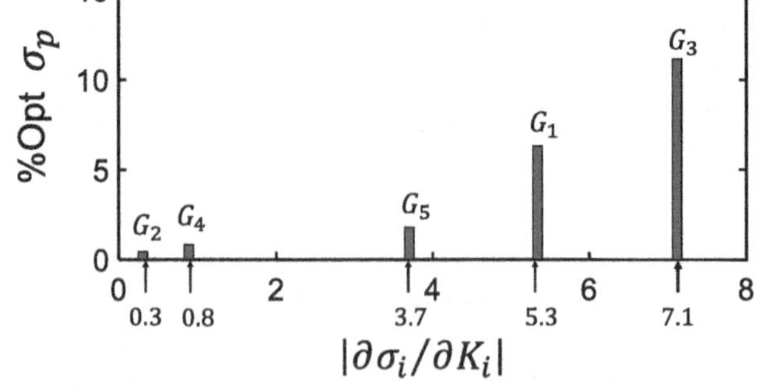

Fig. 5.11 Optimization of a logic path

A more refined strategy to select the best gates to size up should consider the impact on area cost when multi-input logic gates are used in the logic path. For instance, sizing up by a certain amount ΔK an inverter is not the same than in Nand gate that has more than one transistor in the stack of its nMOS network. Sizing up transistors by the same amount ΔK consumes more area in a Nand gate than in an inverter. Also, the previous metric does not consider the interactions of the sized gate with their adjacent gates as shown in Fig. 5.9 and the spatial correlation between gate delays (5.5). For more accuracy, the derivative of the delay standard deviation of the logic path must be evaluated, but this is time-consuming.

Quick Guide for On-Time Design Optimization

Metric (5.19) provides a first approach to select the best gates of a path to size up to reduce the standard deviation of the path delay. The use of this metric requires to pre-characterize the ratio of change of the gate delay standard deviation to changes in the gate size. As this depends mainly on the actual size of the gate in the logic path and its load capacitance, this implies a pre-characterization effort in the design team. Next, a simple way to make a rough evaluation of the metric (5.19) is presented. The nMOS channel width of an inverter is used to represent the gate size (K), and the pMOS network is sized accordingly to keep symmetry. Let us make the partial derivative of the delay standard deviation of an inverter with respect to the channel width of the transistor.

Replacing Eq. (4.8) in Eq. (4.6) and derivating the resultant expression with respect to W, the following expression is obtained:

$$\begin{aligned}\frac{\partial\sigma}{\partial W} &= -\frac{1}{W}\cdot\sqrt{S_{D,W}^2\sigma_W^2 + S_{D,L}^2\sigma_L^2 + S_{D,T_{\mathrm{ox}}}^2\sigma_{D,T_{\mathrm{ox}}}^2 + S_{V_{\mathrm{th}}}^2\sigma_{V_{\mathrm{th}}}^2} \\ &\quad -\frac{1}{W}\cdot\frac{S_{D,W}^2\sigma_W^2}{\sqrt{S_{D,W}^2\sigma_W^2 + S_{D,L}^2\sigma_L^2 + S_{D,T_{\mathrm{ox}}}^2\sigma_{D,T_{\mathrm{ox}}}^2 + S_{V_{\mathrm{th}}}^2\sigma_{V_{\mathrm{th}}}^2}}\end{aligned} \tag{5.20}$$

The first term in Eq. (5.20) dominates over the second. Let us remember that the delay sensitivities reduce as the transistor channel width of the inverter (drive strength) decreases and increase as the load capacitance increases (Eqs. (4.8a)–(4.8d)). Thus, from Eq. (5.20) it can be stated that:

- $\partial\sigma/\partial W$ presents higher scores as the transistor drive strength (channel width) of the inverter has lower values.
- $\partial\sigma/\partial W$ presents higher scores as the load capacitance of the inverter has higher values.

Hint *For reducing the delay standard deviation of a logic path, as a first guess a designer can select to size up those gates with lower values of driven strength and higher values of load capacitance.*

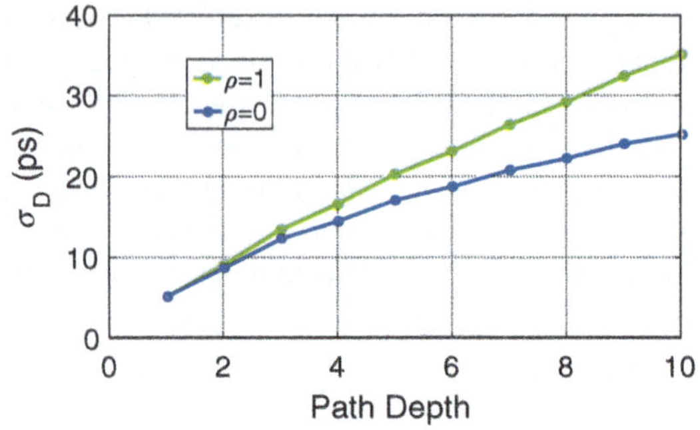

Fig. 5.12 Impact of logic depth on path behavior

Nevertheless, keeping simplifications in mind, this quick guide for on-time design optimization gives the designer a guide on where he should look for reducing the delay standard deviation of a logic path. This guide approximation would bring the benefit of reducing the number of time-consuming Monte Carlo simulations to fulfill time specifications of a circuit.

5.2.4 Logic Depth on Path Delay

5.2.4.1 Impact of Logic Depth on Path Delay Behavior

Equations (5.12) and (5.13) show the strong dependence of the logic depth on the path delay variance. This dependence is illustrated in Fig. 5.12 where the standard deviation of the path delay is plotted as a function of the path logic depth. A fully correlated ($\rho = 1$) case and a non-correlated ($\rho = 0$) case between gates are shown. It is observed that the path delay standard deviation increases for both the fully correlated case and the non-correlated case. However, the impact of the logic depth on the path delay standard deviation is more significant for $\rho = 1$ than for $\rho = 0$. The separation between the two curves increases as the logic depth increases. This separation indicates that for longer paths the impact of correlated variations on path delay is more significant than the impact of pure random variations.

5.2.4.2 Delay Variability of a Logic Path

The delay variability is a metric that allows quantifying how much significant is the path delay standard deviation on the overall path delay. This metric is defined as the ratio of the delay variations of a path (σ_{D_P}) to the nominal delay (μ_{D_P}). Thus, the delay variability is defined as σ_{D_P}/μ_{D_P}. Let us use Eq. (5.12), which is the variance of the path delay for a chain of N inverters. Replacing the covariance between gate delays by the correlation between gate delays (Eq. (2.21)) in Eq. (5.12), the variance of a chain of N inverters can be expressed as [13]:

$$\sigma_{Dp}^2 = \sum_{i=1}^{N}\sum_{j=1}^{N} \sigma_{D_i} \cdot \sigma_{D_j} \cdot \rho_{D_{ij}} \tag{5.21}$$

where σ_{D_i} is the delay distribution of the inverter at location *i*, σ_{D_j} is the delay distribution of the inverter at location *j*, and $\rho_{D_{ij}}$ is the correlation between the delay distributions of the inverters located at *i* and *j*.

Next, the mean (μ_{inv}) and standard deviation (σ_{inv}) of the delay of the inverters are assumed having the same value as they are equally sized and each one drives another similar inverter.

Fully Correlated Case

For the fully correlated case, ρ is equal to 1 for all the terms of the sum in Eq. (5.21). Thus, the variance of the delay at the output of the logic path is given by

$$\sigma_{Dp(\rho=1)}^2 = N^2 \cdot \sigma_{\text{inv}}^2 \tag{5.22}$$

The mean of the delay of the logic path is $\mu_{Dp} = N \cdot \mu_{\text{inv}}$. The delay variability (σ_{DP}/μ_{Dp}) for the fully correlated case is given by Amin et al. [13], Blaauw [17],

$$\left(\frac{\sigma_{Dp}}{\mu_{Dp}}\right)_{\rho=1} = \frac{\sqrt{N^2 \cdot \sigma_{\text{inv}}^2}}{N \cdot \mu_{\text{inv}}} = \left(\frac{\sigma}{\mu}\right)_{\text{inv}} \tag{5.23}$$

Equation (5.23) tells us that the impact of the correlated variations on the overall path delay does not decrease as the logic depth (*N*) increases. The delay variability of the path remains equal to delay variability of the single inverter cell $(\sigma/\mu)_{\text{inv}}$ as the logic depth increases.

Non-correlated Case

For the non-correlated (pure random variations) case, ρ is equal to 0 for all the terms of the sum in Eq. (5.21) except for those terms referring to the correlation of the delay distributions of the same gate where ρ is equal to 1. The variance of the delay at the output of the logic path is given by

$$\sigma_{Dp(\rho=0)}^2 = N \cdot \sigma_{\text{inv}}^2 \tag{5.24}$$

The delay variability (σ_{DP}/μ_{Dp}) for the non-correlated case is given by Amin et al. [13], Blaauw [17],

$$\left(\frac{\sigma_{Dp}}{\mu_{Dp}}\right)_{\rho=0} = \frac{\sqrt{N \cdot \sigma_{\text{inv}}^2}}{N \cdot \mu_{\text{inv}}} = \left(\frac{\sigma}{\mu}\right)_{\text{inv}} \cdot \frac{1}{\sqrt{N}} \tag{5.25}$$

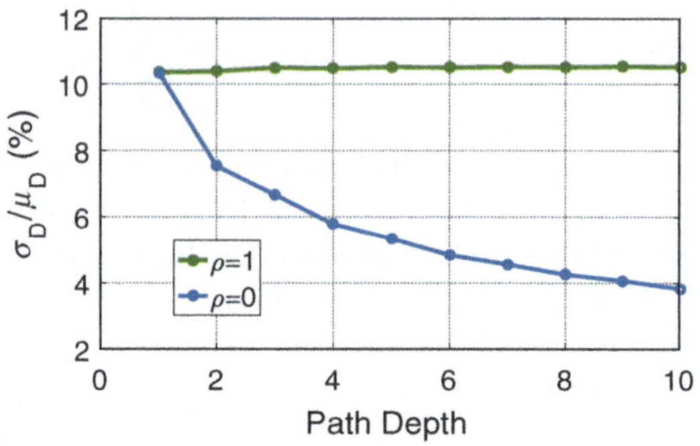

Fig. 5.13 Delay variability of a logic path

Equation (5.25) tells us that the impact of the pure random variations on the overall path delay decreases as the logic depth increases. The delay variability of the inverter cell decreases with the square root of the path logic depth (N).

Figure 5.13 plots the delay variability of a logic path versus the logic depth obtained with SPICE simulations. Two cases are considered: the first case considers that the delay distributions between each pair of inverters are fully correlated, and the second case considers a non-correlated case. Figure 5.13 clearly agrees with the predicted behavior of the theoretical Eqs. (5.23) and (5.25). The impact of correlated variations on the overall path delay does not decrease with logic depth for correlated variations, and the impact of pure random variations on the overall path reduces with logic depth.

Hint *Correlated variations impact more the overall path delay than pure random variations as the path is longer.*

5.3 Statistical Design and Corner Design

Corner-based design assumes pessimistic values for the technological parameters due to process variations during fabrication. For the worst-case corner condition (slow-slow corner or SS corner), the parameters take those extreme values that simultaneously reduce the current capability at all the transistors, and as a consequence, a worst-case circuit delay is obtained.

Parameter taking simultaneously extreme values is a situation unlikely to occur on a fabricated circuit (e.g. on a die) as worst-case corner design ignore correlations [18–22]. A circuit designed to assure 100% functional yield for the unrealistic worst-case circuit delay leads to area and power consumption penalization since the actual correlations between electrical parameters are not considered [18, 19, 22–24]. A more realistic design approach should use statistical design considering spatial correlation between parameters leading to better performance predictions with respect to silicon [19–21, 24].

Worst-case corner analysis and statistical analysis are performed (Fig. 5.14) on a ten-inverter chain with symmetrically minimum-sized inverters. For the worst-case

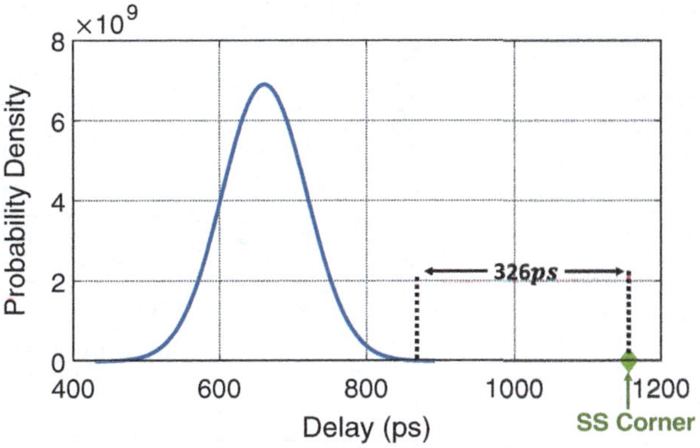

Fig. 5.14 Comparison between statistical design and corner design. Ten-inverter chain with minimum-sized inverters (1×)

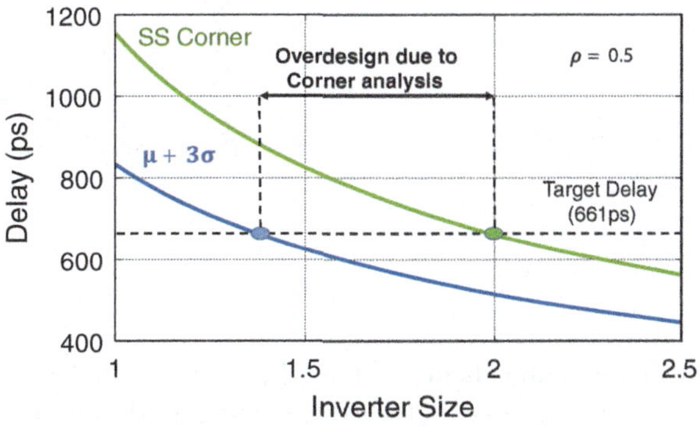

Fig. 5.15 Design of the ten-inverter chain to meet a target delay of 661 ps. $\rho = 0.5$ for the statistical design

corner analysis, the obtained delay at the output of ten-inverter chain at the SS corner is 1157 ps. For the statistical analysis using SPICE Monte Carlo simulations with a spatial correlation value of $\rho = 0.5$ for parameters between inverters, the $\mu + 3\sigma$ value of the statistical delay is 837 ps. The worst-case analysis results in a pessimistic delay of 326 ps larger than the maximum delay obtained with statistical analysis. Thus, a circuit designed to assure 100% functional yield (e.g., all fabricated dies meet the target delay specification) under worst-case analysis would have a larger area and power consumption than a statistical design. In other words, a circuit under worst-case corner is overdesigned [20, 23, 24].

A more detailed analysis of the incurred overhead due to worst-case design is performed next.

Let us assume that a ten-inverter chain must satisfy the target delay of 661 ps (e.g., mean value of the delay distribution shown in Fig. 5.14).

The ten-inverter chain is uniformly sized up until its slowest delay of the distribution ($\mu + 3\sigma$) satisfies the target delay using statistical analysis. In Fig. 5.15 it can be observed that $\mu + 3\sigma$ of the resized chain is satisfied at 1.3× size for all the

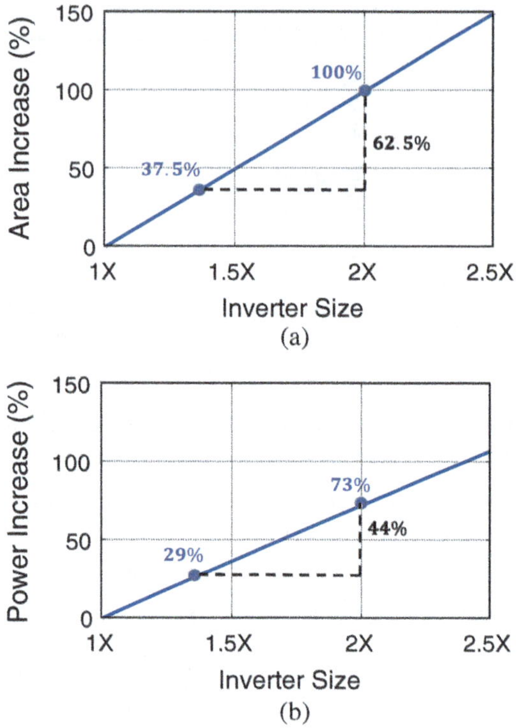

Fig. 5.16 Cost of the ten-inverter chain to meet the target delay. (**a**) Area overdesign. (**b**) Power consumption overdesign

inverters in the chain. On the other hand, a similar sizing-up procedure is followed to satisfy the target delay using worst-case design (Fig. 5.15). In this case, the target delay is satisfied at $2\times$ size of all the inverters in the chain. Hence, a worst-corner design uses larger transistors than a statistical-based design.

The area overhead of the entire ten-inverter chain using corner-based design is 62.5% larger than using a statistical-based design (Fig. 5.16a). The power consumption overhead is also estimated, and for illustration purposes, it has been measured at the nominal values of the resultant distributions. The estimated power consumption overhead of the ten-inverter chain using corner-based design is 44% larger than using a statistical-based design (Fig. 5.16b).

The previous analysis shows the convenience of statistical-based design over corner-based design in order to reduce overdesign.

5.4 Summary

In this chapter, the statistical performance of a logic path under process variations was studied. The spatial correlation due to the distances between gates was considered. A methodology to compute the statistical output delay in a two-inverter chain was described in detail. Then, the methodology is extended to a path composed of different types of gates. The impact of the spatial correlation, relative gate delay sensitivities, sizing a logic path, and logic depth on the delay variance performance of the path was analyzed. An analysis of the cost incurred using corner-based design and a statistical-based design was made.

The path delay variance is determined by (key concepts to remember):

- The delay variances of each gate in the logic path.
- The covariances of the delay distributions between each pair of gates in the logic path.
- The covariance of the delay distributions between a pair of gates depends on their delay sensitivities and their physical distance separation in the layout.

References

1. A. Agarwal, D. Blaauw, V. Zolotov, S. Sundareswaran, M. Zhao, K. Gala, R. Panda, Path-based statistical timing analysis considering inter-and intra-die correlations, in *Proceeding of TAU*, pp. 16–21 (2002)
2. K. Kang, B.C. Paul, K. Roy, Statistical timing analysis using levelized covariance propagation, in *41st Design, Automation and Test in Europe Conference*, Mar 2005, vol. 2, pp. 764–769. https://doi.org/10.1109/DATE.2005.279
3. S. Abbaspour, H. Fatemi, M. Pedram, VGTA: variation-aware gate timing analysis, in *2005 International Conference on Computer Design*, Oct 2005, pp. 351–356. https://doi.org/10.1109/ICCD.2005.115
4. R. Goyal, S. Shrivastava, H. Parameswaran, P. Khurana, Improved first-order parameterized statistical timing analysis for handling slew and capacitance variation, in *20th International Conference on VLSI Design Held Jointly with 6th International Conference on Embedded Systems (VLSID'07)*, Bangalore, Jan 2007, pp. 278–282. https://doi.org/10.1109/VLSID.2007.92
5. M. Orshansky, S. Nassif, D. Boning, *Design for Manufacturability and Statistical Design: A Constructive Approach* (Springer, Boston, 2008)
6. S.S. Sapatnekar, Variability and statistical design. IPSJ Trans. Syst. LSI Des. Methodol. **1**, 18–32 (2008)
7. H. Chang, S.S. Sapatnekar, Statistical timing analysis under spatial correlations. IEEE Trans. Comput. Aided Des. Integr. Circuits Syst. **24**(9), 1467–1482 (2005). https://doi.org/10.1109/TCAD.2005.850834
8. T. Kouno, H. Onodera, Consideration of transition-time variability in statistical timing analysis, in *2006 IEEE International SOC Conference*, Sept 2006, pp. 207–210. https://doi.org/10.1109/SOCC.2006.283882
9. T.T. Soong, *Fundamentals of Probability and Statistics for Engineers* (Wiley, Hoboken, 2004)
10. A. Papoulis, S.U. Pillai, *Probability, Random Variables and Stochastic Processes*, 4th edn. (McGraw-Hill, New York, 2002)
11. M.H. DeGroot, M.J. Schervish, *Probability and Statistics*, 4th edn. (Addison-Wesley, Reading, MA, 2012)
12. C. Forzan, D. Pandini, Statistical static timing analysis: a survey. Integr. VLSI J. **42**(3), 409–435 (2009) . ISSN 0167-9260. Special Section on DCIS2006. https://doi.org/10.1016/j.vlsi.2008.10.002
13. C.S. Amin et al., Statistical static timing analysis: how simple can we get? in *Proceedings 42nd Design Automation Conference*, New York, NY (Association for Computing Machinery, New York, 2005), pp. 652–657. https://doi.org/10.1145/1065579.1065751

14. A. Agarwal, K. Chopra, D. Blaauw, V. Zolotov, Circuit optimization using statistical static timing analysis, in *Proceedings 42nd Design Automation Conference*, June 2005, pp. 321–324 (2005). https://doi.org/10.1109/DATE.2005.281
15. A. Datta, S. Bhunia, S. Mukhopadhyay, K. Roy, A statistical approach to area-constrained yield enhancement for pipelined circuits under parameter variations, in *14th Asian Test Symposium (ATS'05)*, Dec 2005, pp. 170–175. https://doi.org/10.1109/ATS.2005.16
16. V. Champac, A.N.H. Reyes, A.F. Gomez, Circuit performance optimization for local intra-die process variations using a gate selection metric, in *2015 IFIP/IEEE International Conference on Very Large Scale Integration (VLSI-SoC)*, Daejeon, pp. 165–170 (2015). https://doi.org/10.1109/VLSI-SoC.2015.7314410
17. D. Blaauw, K. Chopra, A. Srivastava, L. Scheffer, Statistical timing analysis: from basic principles to state of the art. IEEE Trans. Comput. Aided Des. Integr. Circuits Syst. **27**(4), 589–607 (2008). https://doi.org/10.1109/TCAD.2007.907047
18. H. Chang, S.S. Sapatnekar, Statistical timing analysis considering spatial correlations using a single pert-like traversal, in *ICCAD-2003 International Conference on Computer Aided Design*, Nov 2003, Washington, DC, pp. 621–625. https://doi.org/10.1109/ICCAD.2003.159746
19. B.P. Wong, A. Mittal, Y. Cao, G. Starr, *Nano-CMOS Circuit and Physical Design* (Wiley, Hoboken, 2005). https://doi.org/10.1002/0471653829.ch11
20. X. Li, J. Le, L.T. Pileggi, Statistical performance modeling and optimization. Found. Trends Electron. Des. Autom. **1**(4), 331–480 (2007). https://doi.org/10.1561/1000000008
21. M. Orshansky, S. Nassif, D. Boning, *Design for Manufacturability and Statistical Design: A Constructive Approach* (Springer, Boston, 2008)
22. M. Onabajo, J.G. Silva-Martinez, *Analog Circuit Design for Process Variation-Resilient Systems-on-a-Chip* (Springer, New York, 2012)
23. C. Forzan, D. Pandini, Statistical static timing analysis: a survey. Integr. VLSI J. **42**(3), 409–435 (2009). ISSN 0167-9260. Special Section on DCIS2006. https://doi.org/10.1016/j.vlsi.2008.10.002
24. X. Zhang, X. Bai, Process variability-induced timing failures – a challenge in nanometer CMOS low-power design, in *Emerging Technologies and Circuits*, ed. by A. Amara, T. Ea, M. Belleville. Lecture Notes in Electrical Engineering, vol. 66 (Springer, Dordrecht, 2010). https://doi.org/10.1007/978-90-481-9379-012

Chapter 6
Circuit Analysis Under Process Variations

Abstract The two previous chapters covered analysis at cell and logic path levels. This chapter studies the impact of process variations on the delay performance of some illustrative circuits of general interest. The statistical delay is analyzed based on the methodology described in the previous chapters and also on SPICE Monte Carlo analysis. First, some circuits composed of several logic paths such as 1-bit full adder, 4-bit ripple carry adder, and a benchmark ISCAS circuit are considered. Some measures that the designer may take to improve these designs are discussed. Afterward, the performance of a 6T-SRAM cell memory is analyzed. A high-sigma analysis is also illustrated for this cell, as it is convenient for any cell that can be replicated many times in a chip.

6.1 1-Bit Full Adder

This section gives an overview of the operation of the 1-bit full adder, and then the impact of process variation on its performance is analyzed.

6.1.1 Mirror Adder

A general description of a 1-bit full adder is presented. The adder is an operator widely used in a datapath of digital systems [1–3]. There are many ways to implement an adder [1, 2, 4–6]. The performance characteristics of the adders differ on speed, power, and area constraints, and they are chosen depending on the application to be embedded. A 1-bit full adder is a circuit that makes the sum of three bits: two bits (A and B) correspond to the input data, and the third bit (C) is the input carry that was produced by the preceding adder (Fig. 6.1). The output results of a 1-bit full adder are the sum (S) and an output carry (Co) which are defined by

V. Champac, J. G. Gervacio, *Timing Performance of Nanometer Digital Circuits Under Process Variations*, Frontiers in Electronic Testing 39,
https://doi.org/10.1007/978-3-319-75465-9_6

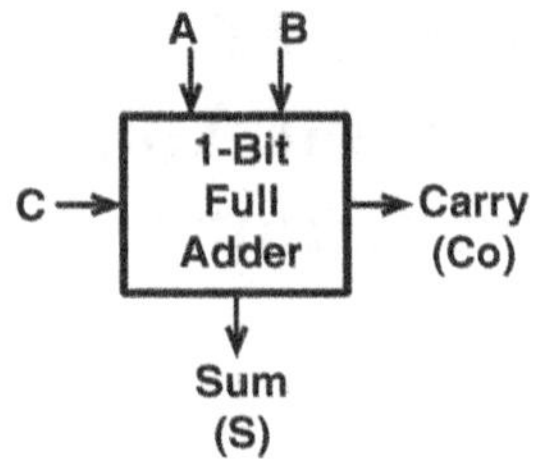

Fig. 6.1 1-bit full adder symbol

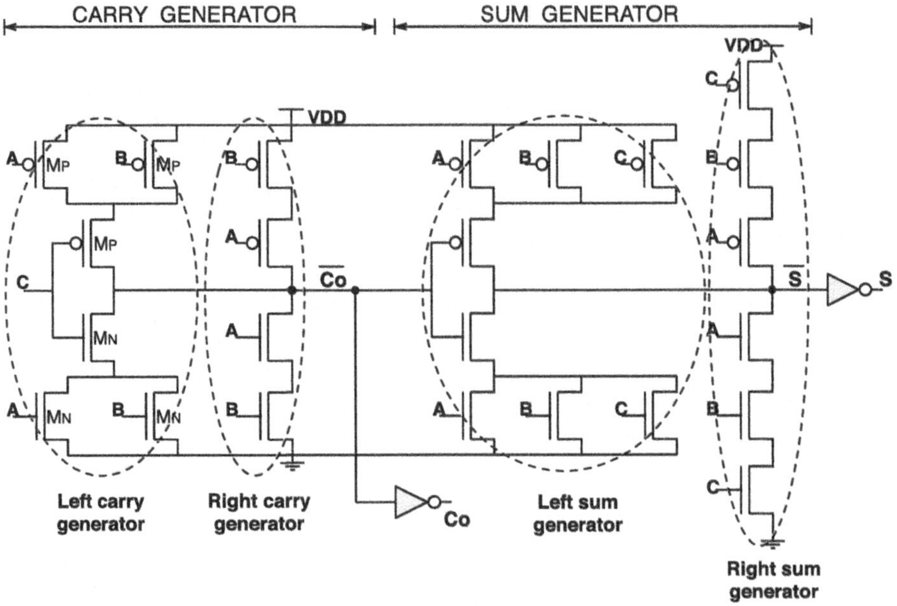

Fig. 6.2 1-Bit mirror adder topology

the Boolean functions $S = (A + B + C)\overline{Co} + ABC$ and $\overline{Co} = (A + B)C + AB$, respectively. The equation for the sum (S) uses the carry that the same 1-bit full adder generates.

Figure 6.2 shows a typical topology of a 1-bit full adder at the transistor level. This topology is known as *mirror adder* since the pMOS network has an identical structure to that of the nMOS network [1, 2]. Two blocks constitute this adder, the carry generator and sum generator blocks. Two segments compose the carry generator block; each one can generate a carry. The left carry generator depends on the input carry, and the right carry generator depends only on the input data. Two segments also compose the sum generator block; each can independently generate a sum. Both segments depend on the three inputs A, B, and C, but the left sum generator additionally depends on the carry produced by the carry generator block ($\overline{Co}$).

The 1-bit adder cell is commonly used in an N-bit ripple carry adder (RCA), which is composed of N full adders connected in series [1, 2]. The carry output of

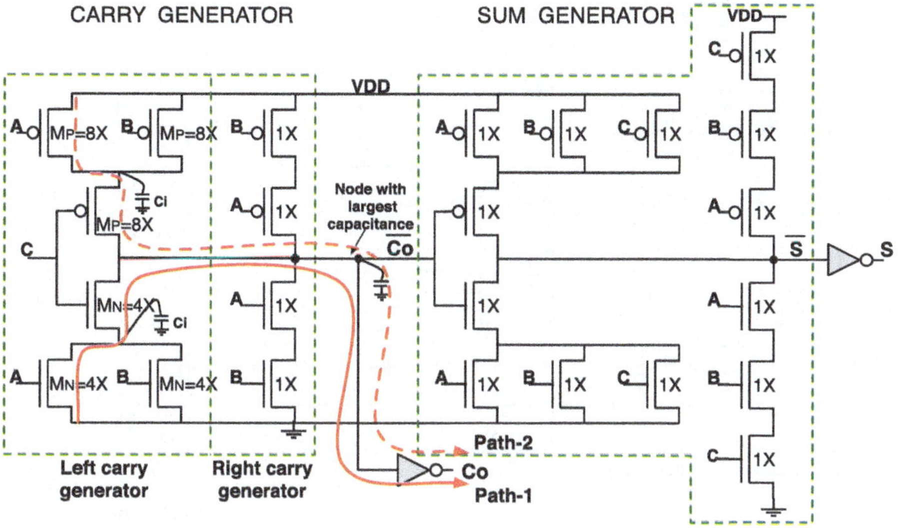

Fig. 6.3 Paths with larger delays of the carry generator block of the 1-bit mirror adder

Table 6.1 Input patterns that activate the critical paths of the mirror adder

Input pattern	ABC	Path activated
P1	000 → 101	Path-1
P2	111 → 010	Path-2
P3	101 → 010	Path-3
P4	010 → 101	Path-4
P5	000 → 111	Path-5
P6	111 → 000	Path-6

one adder is connected to the carry input of the following one. In order to achieve high timing performance of an RCA, the mirror adder is designed with a carry delay as low as possible. The left carry generator that receives an input carry impacts the carry delay of an RCA more than the right carry generator. Because of this, the left carry generator is designed with larger transistors and being careful to minimize its loading capacitance $\overline{Co}$. Figure 6.3 shows the 1-bit mirror adder with its transistor sizes [1] where 1× denotes the minimum-sized transistor allowed by the technology.

6.1.2 Paths in the 1-Bit Full Mirror Adder

The paths of the 1-bit full mirror adder susceptible to exhibit the larger delays are analyzed in this section. Table 6.1 shows the input patterns that activate those paths with larger delays in the 1-bit mirror adder.

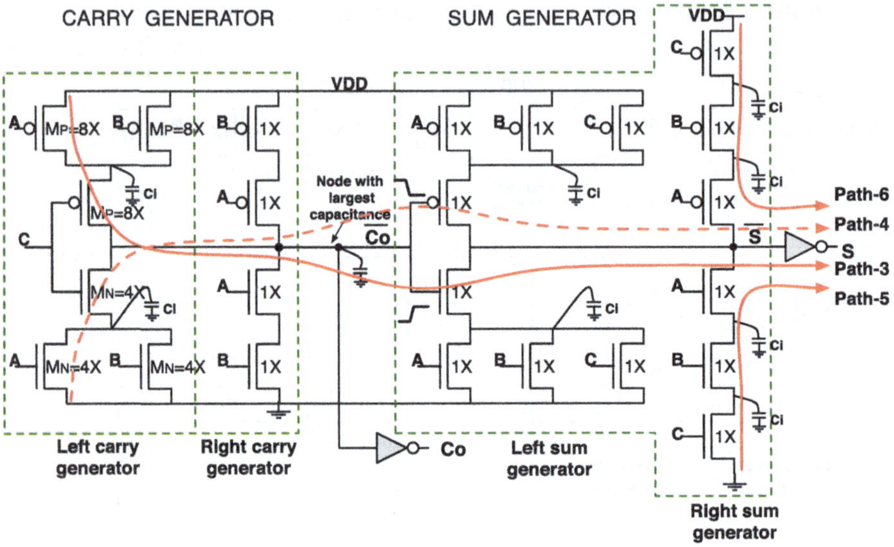

Fig. 6.4 Paths with larger delays of the sum generator block of the 1-bit mirror adder

Figure 6.3 shows the paths with larger delays in the carry generator block. The paths are activated as follows:

- *Path-1* is activated by input pattern P1 (Table 6.1). Pattern P1 activates the nMOS network of the left carry generator. A high-to-low transition occurs at node $\overline{Co}$ that propagates through the output inverter to produce the true carry Co.
- *Path-2* is activated by input pattern P2 (Table 6.1). Pattern P2 activates the pMOS network of the left carry generator. A low-to-high transition occurs at node $\overline{Co}$ that propagates through the output inverter to produce the true carry Co.

Figure 6.4 shows the paths with larger delays in the sum generator block of the mirror adder. The paths are activated as follows:

- *Path-3* is activated by input pattern P3 (Table 6.1). Pattern P3 activates the pMOS network of the left carry generator. A low-to-high transition occurs at node $\overline{Co}$ activating the N-network of the left sum generator. Then, a high-to-low transition occurs in $\overline{S}$ that propagates through the output inverter to produce the true sum S.
- *Path-4* is activated by input pattern P4 (Table 6.1). Pattern P4 activates the nMOS network of the left carry generator. A high-to-low transition occurs at node $\overline{Co}$ activating the P-network of the left sum generator. Then, a low-to-high transition occurs in $\overline{S}$ that propagates through the output inverter to produce the true sum S.
- Path-5 and Path-6 are activated by input patterns P5 and P6 (Table 6.1), respectively. Path-5 and Path-6 are created when all the signals A, B, and C switch simultaneously.

Table 6.2 Standard deviation of technological parameters

Standard deviation	Nmos	Pmos
σ_W (nm)	6	6
σ_L (nm)	3	3
$\sigma_{T_{ox}}$ (nm)	0.119	0.121
$\sigma_{V_{th}}$ (mV)	36.7	35.8

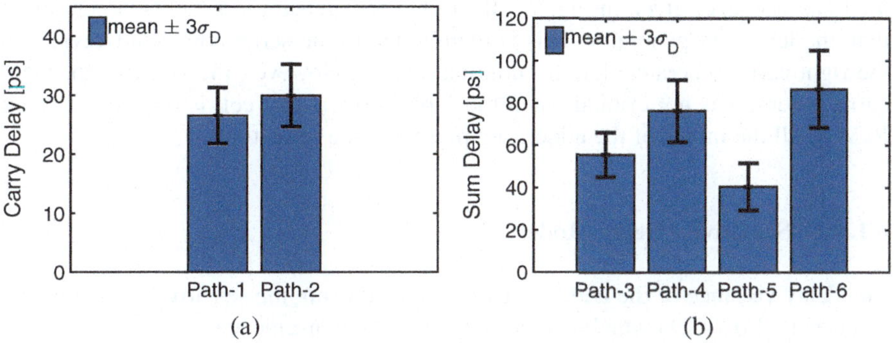

Fig. 6.5 Statistical delay performance of the main paths of the 1-bit mirror adder using SPICE Monte Carlo simulations. (**a**) Carry generator block. (**b**) Sum generator block

6.1.3 *Statistical Delay*

The performance of the mirror adder under process variations is analyzed using both SPICE Monte Carlo simulations and the sensitivity delay model. Correlated variations are considered for L, W, and T_{ox} and purely random variations for V_{th}. Parameters L, W, and T_{ox} are fully correlated due to the short distance separation between transistors in a single adder cell. Table 6.2 shows the standard deviations for the parameters W, L, T_{ox}, and V_{th}. The value of the standard deviation of the threshold voltage shown in Table 6.2 corresponds to a transistor with minimum dimensions. The transistor threshold voltages accounting for the effect of RDF are calculated using Eq. (3.2).

6.1.3.1 Monte Carlo Analysis

Figure 6.5 shows a bar graph of the statistical delay performance of the paths with larger delays of the carry generator block and the sum generator block using SPICE [7]. The top of the bars gives the mean value (μ_D) of the path, and the inner vertical line in the bar gives the spreading of the delay ($\pm 3\sigma_D$).

The path delays in the sum generator block are larger than those in the carry generator block. Path-2, which is activated with the input pattern P2 (Table 6.1), presents the largest delay for the carry generator block. The maximum delay of Path-2 is $\mu_D + 3\sigma_D = 35.22$ ps. On the other hand, for the sum generator block,

Path-4 and Path-6 present the larger delays. The maximum delay of Path-4 that is activated with input pattern P4 is $\mu_D + 3\sigma_D = 91$ ps, and the maximum delay of Path-6 that is activated with input pattern P6 is $\mu_D + 3\sigma_D = 104.9$ ps.

In spite that the delays of the sum generator block are larger than those of the carry generator block, the main concern is the delay response of the carry generator block when the 1-bit mirror adder is used in an N-bit RCA. For the mirror adder with the transistor sizes given in Fig. 6.3, the right carry generator presents larger delays than the left carry generator. This is mainly due to the serial-connected devices in the right carry generator that are minimum sized. However, the delay of the right carry generator is not critical when the 1-bit mirror adder cell is used in an N-bit RCA as all the inputs of the adders are present at the same time.

6.1.3.2 Sensitivity Delay Model

The delay variance of the adder is also computed using the sensitivity delay model as given by Eq. (4.21) which is repeated below for convenience,

$$\sigma_D^2 = S_{D,W}^2\sigma_W^2 + S_{D,L}^2\sigma_L^2 + S_{D,T_{\text{ox}}}^2\sigma_{T_{\text{ox}}}^2 + \sum_{i=1}^{p} S_{D,V_{\text{th}_i}}\sigma_{V\text{th}_i} \tag{6.1}$$

where the first three terms are due to correlated variations and the last term is due to purely random variations. $S_{D,L}$, $S_{D,W}$, and $S_{D,T_{\text{ox}}}$ are delay sensitivities to the variation of the parameters L, W, and T_{ox}, respectively. $S_{D,V_{\text{th}_i}}$ is the delay sensitivity of the i-transistor to the variation of V_{th}. The delay sensitivities depend on the input pattern. σ_L, σ_W, and $\sigma_{T_{\text{ox}}}$ are the standard deviations of the parameters. More information about the methodology of delay computation at cell level can be found in Chap. 4.

Table 6.3 shows the delay standard deviation (σ) of the paths with larger delays in the 1-bit mirror adder obtained with Eq. (6.1). The delay behavior of the adder using delay sensitivities is similar to that observed with SPICE. The results obtained with Eq. (6.1) (Table 6.3) agree with those obtained using SPICE (Fig. 6.5). The average error of estimation of the standard deviation of the path delay between SPICE and Eq. (6.1) is of around 3.8%.

Table 6.3 Delay deviation of the main paths of the 1-bit mirror adder using precomputed delay sensitivities

Path	σ_D (ps)
Path-1	1.53
Path-2	1.7
Path-3	3.32
Path-4	4.63
Path-5	3.62
Path-6	5.93

Hint *Effort needs to be devoted to optimize the critical paths in the carry generator block when the single adder cell is planned to be used in an N-bit ripple carry adder.*

6.2 4-Bit Ripple Carry Adder

A ripple carry adder (RCA) is a circuit that performs the sum of two binary numbers [1, 2]. An N-bit RCA (Fig. 6.6) is implemented by connecting N full adders in series; the carry output of one adder is connected to the carry input of the following one. The first adder of the chain receives the initial carry. Each full adder generates a carry, which is used to generate the sum of the next adder of the chain. The output bits of the sum are processed and available after the corresponding carry input bits have arrived at the adder.

6.2.1 Critical Logic Paths of a 4-Bit RCA

Logic paths with the larger delays are known as *critical paths*. Figure 6.7 shows the critical paths of the 4-bit ripple carry adder. Path-1 starts at the least significant input bits (A1, B1) and propagates through the carry generator blocks of all the adders, except the carry stage of the last adder in the chain, until reaching the most significant output bit of the sum (S4). Path-2 starts at the input carry of the adder (Ci) and propagates through the carry generator blocks of all the adders until reaching the output carry of the RCA (C4).

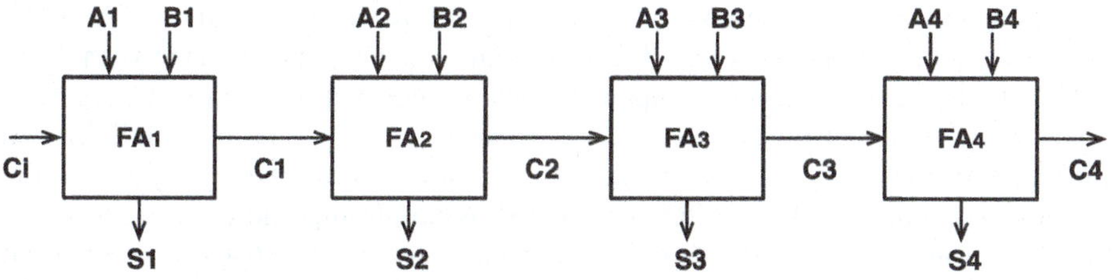

Fig. 6.6 4-bit ripple carry adder. FA, full adder

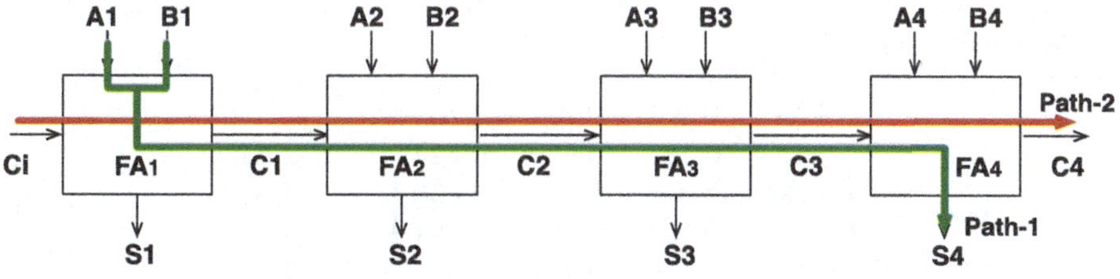

Fig. 6.7 Critical paths of the 4-bit ripple carry adder

The total propagation delay of Path-1 in an N-bit RCA is given by summing up the carry delays of all the adders in the chain, except for the last one, with the sum delay of the last adder in the chain. The worst-case delay for Path-1 is given by

$$T_{\text{Path-1}} = (N - 1)T_{\text{carry}} + T_{\text{sum}} \tag{6.2}$$

where T_{carry} is the propagation delay to generate the carry of a single full adder and T_{sum} is the propagation delay to generate the sum of a single full adder.

The total propagation delay of Path-2 in an N-bit RCA is given by summing up the carry delays of all the adders. The worst-case delay for Path-2 is given by

$$T_{\text{Path-2}} = NT_{\text{carry}} \tag{6.3}$$

According to Eqs. (6.2) and (6.3), the propagation delay of the N-bit RCA is linearly proportional to the number of bits (N).

6.2.2 Statistical Delay

The delay performance of ripple carry adders is impacted by process variations of the parameters [8–11]. A 4-bit RCA using a mirror adder cell is studied using both SPICE Monte Carlo simulations and the sensitivity delay model. Correlated variations in parameters L, W, and T_{ox} and purely random variations of V_{th} are considered. Table 6.2 shows the standard deviation of the process parameters. The transistor threshold voltages accounting for the effect of RDF are calculated using Eq. (3.2).

The spatial correlation between parameters needs to be considered in a 4-bit RCA as the adder cells composing the adder have some distance separation between them. The formal methodology to compute the spatial correlations between parameter uses the exponential model described in Chap. 3. For the sake of simplicity and to maintain the main focus of the analysis on adder performance evaluation, the values of spatial correlations are approached assuming that the adder cells of the 4-bit RCA are placed horizontally one after the other. Matrix (6.4) shows the used spatial correlation values between pair of adder cells (Fig. 6.6). For instance, the correlation between adder FA_1 and adder FA_2 is 0.78, the correlation between adder FA_1 and adder FA_3 is 0.63, and so forth. A correlation value of 1 means that it refers to the same adder.

$$\rho_{\text{RCA}} = \begin{array}{c} \begin{matrix} FA_1 & FA_2 & FA_3 & FA_4 \end{matrix} \\ \begin{pmatrix} 1 & 0.781 & 0.630 & 0.527 \\ 0.781 & 1 & 0.781 & 0.630 \\ 0.630 & 0.781 & 1 & 0.781 \\ 0.527 & 0.630 & 0.781 & 1 \end{pmatrix} \end{array} \begin{matrix} FA_1 \\ FA_2 \\ FA_3 \\ FA_4 \end{matrix} \tag{6.4}$$

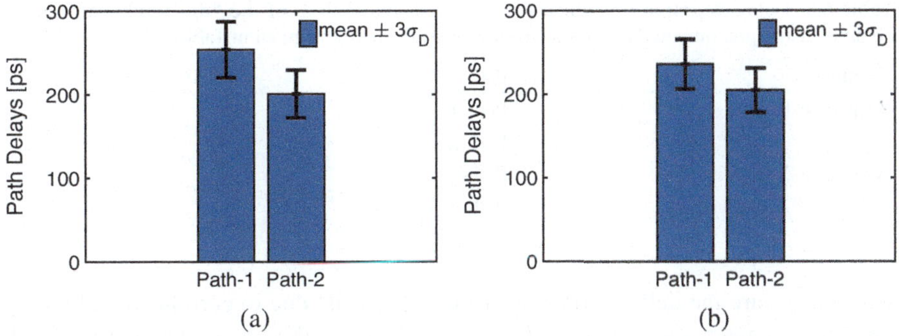

Fig. 6.8 Statistical delay performance of the two critical paths of the 4-bit RCA obtained with SPICE. (**a**) Sequence-1. (**b**) Sequence-2

Table 6.4 Input patterns for the 4-bit RCA

Sequence	$[A_3A_2A_1A_0]$	$[B_3B_2B_1B_0]$	Input carry (Ci)
Sequence-1	[0000] → [1111]	[1111] → [0000]	0 → 1
Sequence-2	[1111] → [0000]	[1111] → [1111]	1 → 0

6.2.2.1 Monte Carlo Analysis

Figure 6.8 shows a bar graph of the statistical delay performance of the two critical paths Path-1 and Path-2 (Fig. 6.7) of the 4-bit RCA using Monte Carlo simulations (SPICE) [7]. The 4-bit RCA is analyzed for the two sequences of input vectors shown in Table 6.4. The top of the bars (Fig. 6.8) gives the mean value (μ_D) of the path, and the inner vertical line in the bar gives the spreading of the delay ($\pm 3\sigma_D$). Critical path Path-1 presents larger delays than critical path Path-2 for the two applied sequences of input vectors. The maximum delays of Path-1 including process variations of the parameters are $\mu_D + 3\sigma_D = 287.32$ ps and $\mu_D + 3\sigma_D = 265.31$ ps for sequence-1 and sequence-2, respectively. Even if the delays of Path-1 are larger than those of Path-2, the statistical performance of both critical paths of the 4-bit RCA needs to be verified due to the overlapping of delays between Path-1 and Path-2 (Fig. 6.8). This overlapping may lead to fabricated dies with delays of Path-2 larger than the delays of Path-1.

6.2.2.2 Sensitivity Delay Model

The delay variance of the 4-bit RCA is also computed using the delay sensitivity model as given by Eq. (4.21), which is slightly rewritten as

$$\sigma^2_{D_{\mathrm{RCA}}} = \sigma^2_{D_{\mathrm{FA}_1}} + \sigma^2_{D_{\mathrm{FA}_2}} + \sigma^2_{D_{\mathrm{FA}_3}} + \sigma^2_{D_{\mathrm{FA}_4}} + 2\sum_{i=1}^{3}\sum_{j=i+1}^{4} \mathrm{Cov}(D_{\mathrm{FA}_i}, D_{\mathrm{FA}_j}) \qquad (6.5)$$

Table 6.5 Delay deviation of critical paths, Path-1 and Path-2, of the 4-bit ripple carry adder obtained using precomputed delay sensitivities with the patterns given in Table 6.4

Sequence	Path	σ_D (ps)
Sequence-1	Path-1	9.89
	Path-2	8.58
Sequence-2	Path-1	8.78
	Path-2	8.01

where $\sigma^2_{D_{\mathrm{FA}}}$ are the delay variances of the adder cells due to correlated and purely random parameter variations and $\mathrm{Cov}(D_{\mathrm{FA}_i}, D_{\mathrm{FA}_j})$ is the covariance between each pair of adders. The variance of each adder is obtained as described in Sect. 6.1.3.

The covariances are computed by

$$\mathrm{Cov}(D_{\mathrm{FA}_i}, D_{\mathrm{FA}_j}) = \sum_{m=1}^{k} S_{Di,m} \cdot S_{Dj,m} \cdot \rho_{i,j} \cdot \sigma_m^2 \tag{6.6}$$

where $S_{Di,m}$ and $S_{Dj,m}$ are the delay sensitivities of the adders i and j, respectively, to the variation of the parameter m, σ_m is the standard deviation of the parameter m, and $\rho_{i,j}$ is the spatial correlation between adders i and j. Correlation values are given in Matrix (6.4).

Table 6.5 shows the delay standard deviation (σ) of the critical paths Path-1 and Path-2 for the two sequences of input vectors using the delay sensitivity model. The delay behavior of the 4-bit RCA using delay sensitivities is similar to that obtained with Monte Carlo simulations. The results obtained with Eq. (6.5) (Table 6.5) agree with those obtained using Monte Carlo simulations (Fig. 6.8). The estimated error between SPICE and the delay sensitivity model is not higher than 8%.

Hint *The performance of the ripple carry adder is determined by its critical logic paths. Therefore, adder optimization should focus on these critical paths.*

6.2.3 Improving Adder Performance

The main disadvantage of the RCA is the propagation of the signal from the adder cell of the least significant bit through all the adders until reaching the adder cell of the most significant bit. This propagation of the signal (ripple) causes a significant delay. Particular attention must be paid to the design, place, and route of the adder cells and interconnects along the critical paths. Devices in the trajectories of the critical paths (Fig. 6.7) may be resized for improving adder performance.

Improvement of adder performance is discussed next. We do not pretend to show an ideal optimization methodology of the adder, but just a glimpse of improvement

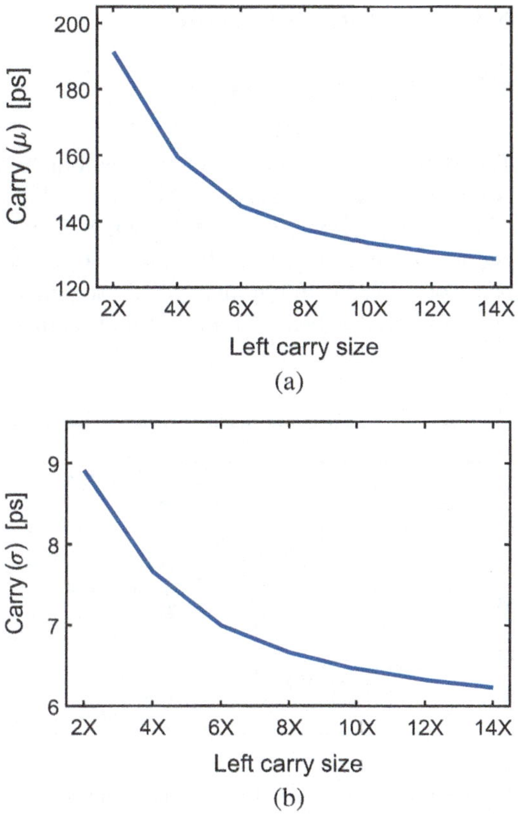

Fig. 6.9 Delay performance of Path-2 of the 4-bit RCA as a function of the channel width of transistors (M_N and $M_P = 2M_N$) located in the left carry generator of the mirror adder. (**a**) Mean. (**b**) Standard deviation

of adder performance by gate-sizing. Figure 6.9 shows the statistical performance of the critical path Path-2 of the 4-bit RCA (Fig. 6.7) as a function of the size of the nMOS transistor in the left carry generator of each adder (Fig. 6.3). The nMOS transistors in the left carry generator block are sized X times the minimum transistor. Also, the inverter gate (Fig. 6.3) producing the carry output (Co) of each adder cell is resized accordingly. The pMOS transistors are resized to keep symmetrical behavior in the left carry generator. Sequence-1 of input vectors (Table 6.4) is used here.

It is observed that both the mean carry and its standard deviation are reduced as the nMOS transistor of the left carry generator is sized up. The rate of reduction of μ_D and σ_D slows down for larger transistors and is very low for a sufficiently large transistor. Note that resizing a gate adds capacitance to the output node, also loading the preceding gate.

Hint *The delay performance of a 4-bit RCA can be improved by properly sizing up the transistors in the left carry generator of the single adder cell.*

6.3 Benchmark Circuit

6.3.1 ISCAS Circuit S27

The statistical performance of the combinational block of ISCAS'89 benchmark circuit S27 [12] is analyzed in this section. Figure 6.10 shows the schematic circuit of the combinational block of circuit S27, whereas Fig. 6.11 illustrates the paths ending at output O1. Usually, static timing analysis is applied to find the critical paths of a circuit. In our analysis, those paths with a logic depth greater than four are considered critical. They are Path-1, Path-2, Path-3, Path-4, and Path-5.

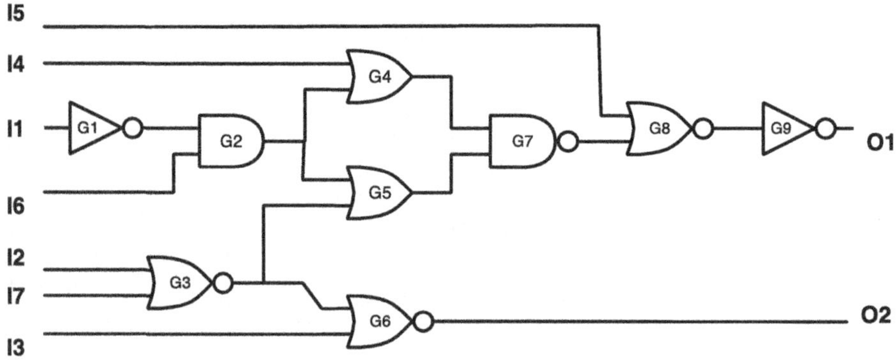

Fig. 6.10 Benchmark circuit S27. Combinational block

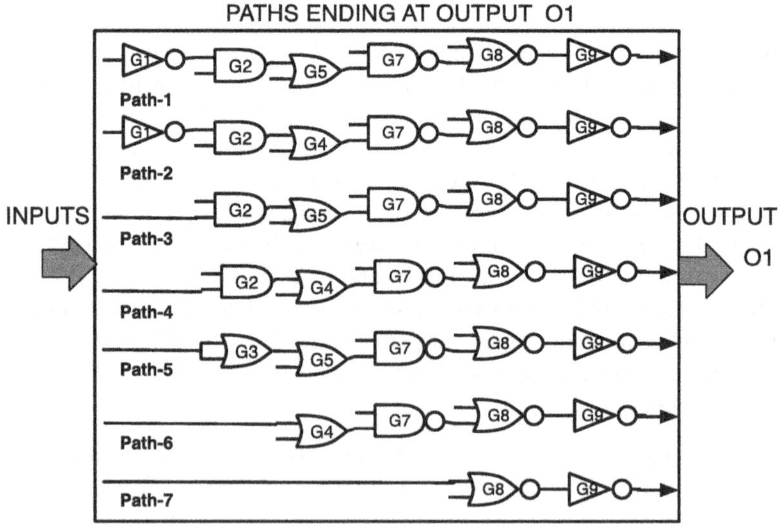

Fig. 6.11 Paths of circuit S27 ending at output O1

6.3.2 *Statistical Delay*

The statistical delay of circuit S27 is obtained using both SPICE Monte Carlo simulations and the sensitivity delay model. Correlated variations in parameters L, W, and T_{ox} and purely random variations in V_{th} are considered. Table 6.2 shows the standard deviation of the process parameters. The transistor threshold voltages accounting for the effect of RDF are calculated using Eq. (3.2). For the purpose of simplicity and to keep the main focus on performance evaluation, the spatial correlation is not estimated from a circuit layout. Matrix (6.4) shows the spatial correlation values, which were approximated, between pairs of logic cells.

$$\rho_{ij} = \begin{array}{c} \begin{array}{ccccccccc} G1 & G2 & G3 & G4 & G5 & G6 & G7 & G8 & G9 \end{array} \\ \begin{pmatrix} 1 & 0.845 & 0.834 & 0.724 & 0.724 & 0.72 & 0.63 & 0.557 & 0.50 \\ 0.845 & 1 & 0.933 & 0.844 & 0.844 & 0.834 & 0.724 & 0.63 & 0.557 \\ 0.834 & 0.933 & 1 & 0.829 & 0.839 & 0.845 & 0.72 & 0.628 & 0.556 \\ 0.724 & 0.844 & 0.829 & 1 & 0.965 & 0.917 & 0.844 & 0.724 & 0.63 \\ 0.724 & 0.844 & 0.839 & 0.965 & 1 & 0.949 & 0.844 & 0.724 & 0.63 \\ 0.72 & 0.834 & 0.845 & 0.917 & 0.949 & 1 & 0.834 & 0.72 & 0.628 \\ 0.63 & 0.724 & 0.72 & 0.844 & 0.844 & 0.834 & 1 & 0.845 & 0.724 \\ 0.55 & 0.63 & 0.628 & 0.724 & 0.724 & 0.72 & 0.845 & 1 & 0.845 \\ 0.50 & 0.557 & 0.556 & 0.63 & 0.63 & 0.628 & 0.724 & 0.845 & 1 \end{pmatrix} \end{array} \begin{array}{c} \\ G1 \\ G2 \\ G3 \\ G4 \\ G5 \\ G6 \\ G7 \\ G8 \\ G9 \end{array} \tag{6.7}$$

6.3.2.1 Monte Carlo Analysis

Spatial correlations between logic gates are taken into consideration using the simulation methodology shown in Appendix C. Figure 6.12 shows a bar graph of the statistical delay performance of the critical paths of circuit S27 (Fig. 6.10) using SPICE Monte Carlo simulations [7]. The top of the bars in Fig. 6.12 gives the mean value (μ_D) of the path, and the inner vertical line in the bar gives the spreading

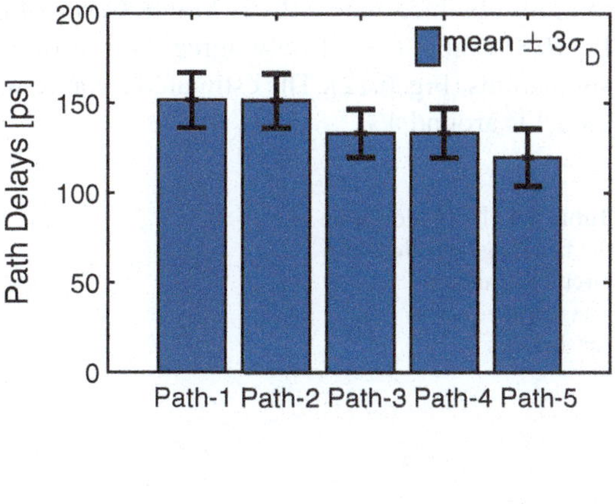

Fig. 6.12 Statistical delay performance of the critical paths ending at output O1 of circuit S27 using SPICE simulations

of the delay ($\pm 3\sigma_D$). Figure 6.12 shows that Path-1 and Path-2 have higher mean delay values than Path-3 and Path-4 and Path-5 has the smallest delay of the set of critical paths. An interesting observation is the overlapping of delays between paths due to the spreading of delays caused by process variations. Path-1 and Path-2 have overlapping delays with Path-3 and Path-4. Even more, larger delays of Path-5 are closer to the smaller delay of the other paths. The previous results show that the delay of the set of critical paths could have values not so much difference among them. Resizing a gate during on-time design optimization may affect the delays of the other paths. Usually, a small or moderate set of critical paths need to be resized during on-time design optimization.

6.3.2.2 Sensitivity Delay Model

The delay variance of circuit S27 is also computed using delay sensitivities as given by Eq. (4.21) which is repeated below for convenience,

$$\sigma_{D_{cp}}^2 = \sum_{i=1}^{N} \sigma_{Di}^2 + \sum_{i=1}^{N-1} \sum_{j=i+1}^{N} \mathrm{Cov}(D_i, D_j) \tag{6.8}$$

where the first term represents the gate delay variances due to correlated and purely random parameter variations and the second term represents the covariances between each pair of gates due to spatial correlation.

The covariances are computed by

$$\mathrm{Cov}(D_i, D_j) = \sum_{m=1}^{k} S_{Di,m} \cdot S_{Dj,m} \cdot \rho_{i,j} \cdot \sigma_m^2 \tag{6.9}$$

where $S_{Di,m}$ and $S_{Dj,m}$ are the delay sensitivities to a parameter m of the gates i and j, respectively, σ_m is the standard deviation of the parameter m and $\rho_{i,j}$ is the spatial correlation between gates i and j.

Table 6.6 shows the delay standard deviation of the critical paths using delay sensitivities. The delay behavior of the circuit S27 using delay sensitivities is similar to those obtained with SPICE Monte Carlo simulations. The results obtained with Eq. (6.8) (Table 6.6) closely agree with those obtained using SPICE Monte Carlo simulations (Fig. 6.12). The estimated error between SPICE and the delay sensitivity model is around 4%.

Table 6.6 Delay deviation of the five longest paths of circuit S27 using precomputed delay sensitivities

Path	Std. deviation (σ_D) (ps)
Path-1	4.95
Path-2	4.85
Path-3	4.29
Path-4	4.41
Path-5	3.83

Hint *A good design practice strategy for on-time design optimization is to identify and enumerate the critical paths of the circuit from the path delay with the largest delay at the top to the path with the smallest delay at the bottom. The initial resizing effort to comply with specifications should go from top to bottom.*

Hint *When improving the performance of a specific logic path by gate resizing, the statistical delay performance of the other critical paths should be under surveillance as they may increase their delay due to loading effects.*

6.3.3 Improving Delay Performance

The use of CAD tools to design optimization of the overall circuit delay has been proposed in the literature [13–15]. The benefits of these strategies are hereafter illustrated. Figure 6.13 plots the standard deviation of the delay of circuit S27 versus the consumed area. Figure 6.13 is constructed by making extensive SPICE simulations. Each dot corresponds to a different sizing condition of the logic gates of circuit S27. The standard delay deviation of the most critical path and the consumed circuit area are measured. It is observed that a target standard deviation of the circuit delay can be determined at different area cost. Furthermore, a desired target standard deviation can be obtained at a minimum area cost. The solid curve gives the optimum designs for minimizing the standard deviation of the circuit at a constrained area cost. The optimum designs can be obtained using sensitivity-guided metrics to select the best gate to size up as well as optimization algorithms [13–15]. Figure 6.13 shows the principal role that circuit sizing has on the statistical response of the circuit, but global design optimization techniques are time consuming, and they present complexity for being used at on-time design optimization. Ad hoc strategies for resizing gates of a logic path are more suitable for on-time design optimization. One possible design strategy was explored in Sect. 5.2.3 of Chap. 5.

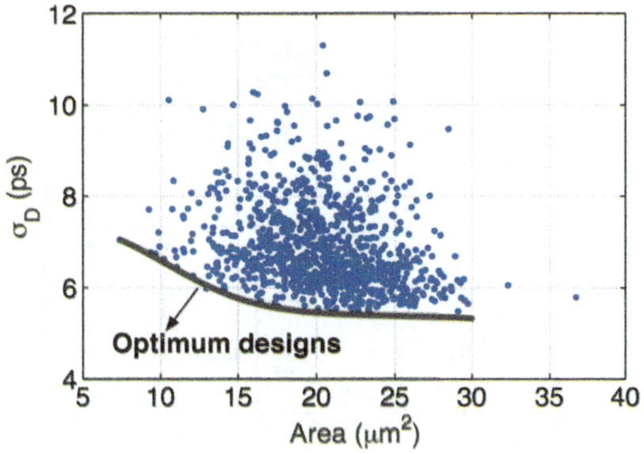

Fig. 6.13 Standard delay deviation as a function of the area cost of circuit S27

6.4 SRAM Cell Memory

A static random-access memory (SRAM) architecture consists of an array of memory cells and peripheral circuits to perform reading and writing operations in the cells [16–19]. Figure 6.14 shows the basic architecture of a SRAM memory. The main blocks composing the SRAM memory are the cell memory array, the precharge circuit, the sense amplifier, the write driver, and the row and column decoder.

6.4.1 6T-SRAM Cell Operation

The memory cell is the more critical element of an SRAM architecture. Figure 6.15 depicts the widely used 6T-SRAM memory cell. The 6T-SRAM cell consists of two

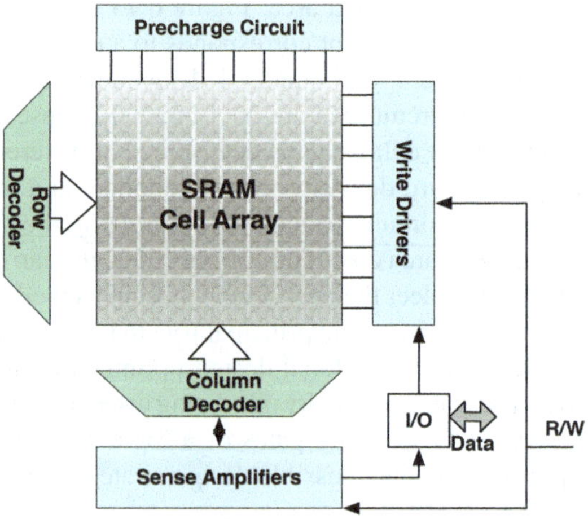

Fig. 6.14 Basic architecture of a SRAM memory

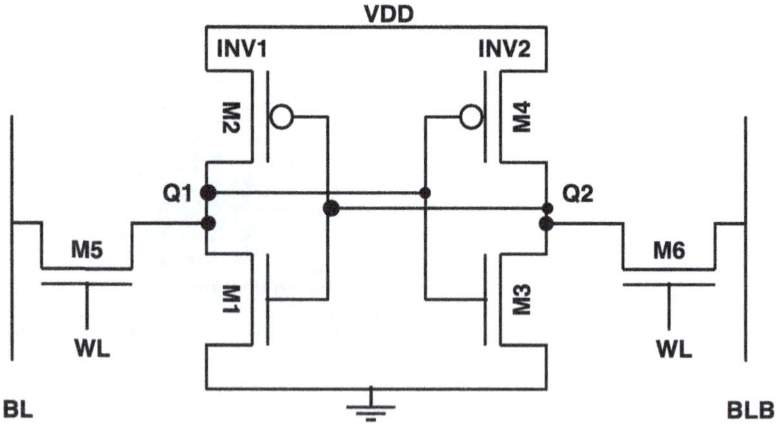

Fig. 6.15 Topology of the 6T-SRAM memory cell

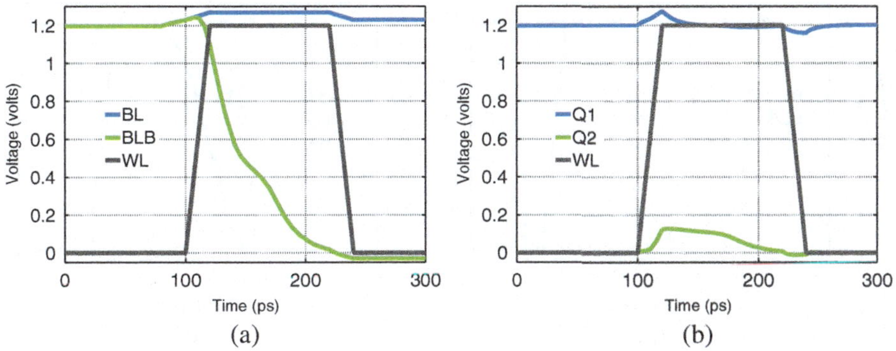

Fig. 6.16 Timing response of a reading operation of the SRAM cell. (**a**) BL and BLB. (**b**) Q1 and Q2

cross-coupled CMOS inverters and two nMOS transistors used as access transistors. The cross-coupled inverters hold the bit data, and the two nMOS transistors provide access to the data. The word line (WL) signal activates the cell for reading and writing operations by means of the transistors M5 and M6. The bit lines BL and BLB controlled by the precharge circuit put the voltage conditions at the access transistors (M5 and M6) for the reading and writing operations. The 6T-SRAM cell must be reliable for the reading and writing operations and holding of the data for as long as the cell is powered up [16–19]. Moreover, these characteristics must be achieved at minimum cell area to allow high-density integration.

6.4.1.1 Reading

Figure 6.16 illustrates the behavior of some signals of the memory cell during a reading operation. The word line (WL), bit line (BL and BLB), and internal data (Q1 and Q2) signals are shown. BL and BLB initially have a high logic value (VDD), and BLB pulls down when the WL signal goes from low to high. Q1 and Q2 hold their logic values in the flip-flop. Observe that Q2 rises slightly due to the discharge of BLB through M6 and M3. The differential voltage between the bit lines, required to perform a correct read operation, depends on the kind of the sense amplifier.

For the *read* operation, the size of the access transistor (M6) is made smaller than the pull-down transistor (M3). $\beta_{M3M6} = \frac{\beta_{M3}}{\beta_{M6}} > 1$ is usually used to ensure correct operation [16, 17, 19].

6.4.1.2 Writing

Figure 6.17 illustrates the behavior of some signals of the memory cell during a writing operation. The word line (WL), bit line (BL and BLB), and internal data (Q1 and Q2) signals are shown. BL and BLB initially have a high logic value (VDD),

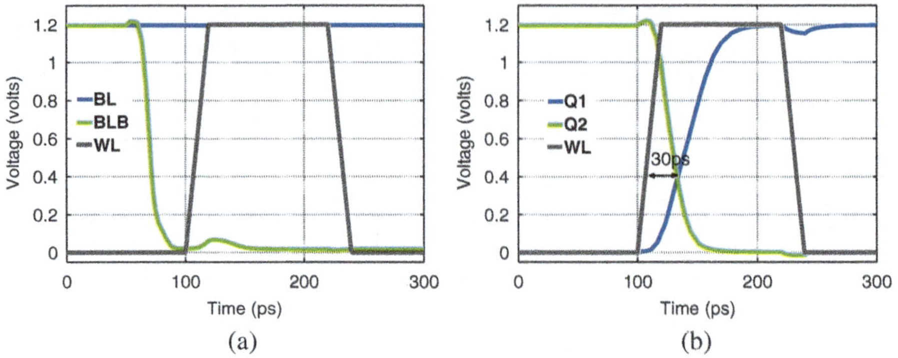

Fig. 6.17 Timing response of a writing operation of the SRAM cell. (**a**) BL and BLB. (**b**) Q1 and Q2

and when the driver is activated, BL remains at VDD, and BLB pulls down. Q1 and Q2 initially have a 0 and 1 logic value, respectively; when the WL signal goes from low to high, Q1 rises toward the switching threshold voltage of the inverter, and Q2 falls toward the same threshold voltage. When either Q1 or Q2 reaches the switching threshold voltage, positive feedback takes place. This process writes a logic 0 in Q2 and a logic 1 in Q1.

For the *write* operation, the size of the access transistor (M5) is made larger than the pull-up transistor (M2). $\beta_{M5M2} = \frac{\beta_{M5}}{\beta_{M2}} > 1$ is usually used to ensure correct operation [16, 17, 19].

6.4.2 *Impact of Process Parameter Variation on the 6T-SRAM Cell Operation*

Process variations of the parameters impact the performance of the SRAM cell [16, 20, 21]. For proper operation of the cell, it is important that the cross-coupled inverters, composed of nMOS and pMOS transistors, are matched in all their parameters [22]. A 6T-SRAM cell is studied using SPICE Monte Carlo simulations. Correlated variations in parameters L, W, and T_{ox} and purely random variations in V_{th} are considered. The standard deviation of the parameters is shown in Table 6.2.

Figure 6.18 illustrates the timing response of the reading and writing operations of the memory cell under process variations. At a reading operation, BLB may have a slower transition due to process variation increasing the time for a correct read operation (Fig. 6.18a). At a writing operation, nodes Q1 and Q2 may increase their delays to switch to the desired voltage values due to process variations (Fig. 6.18b). The increase of the delays at Q1 and Q2 may increase the writing time.

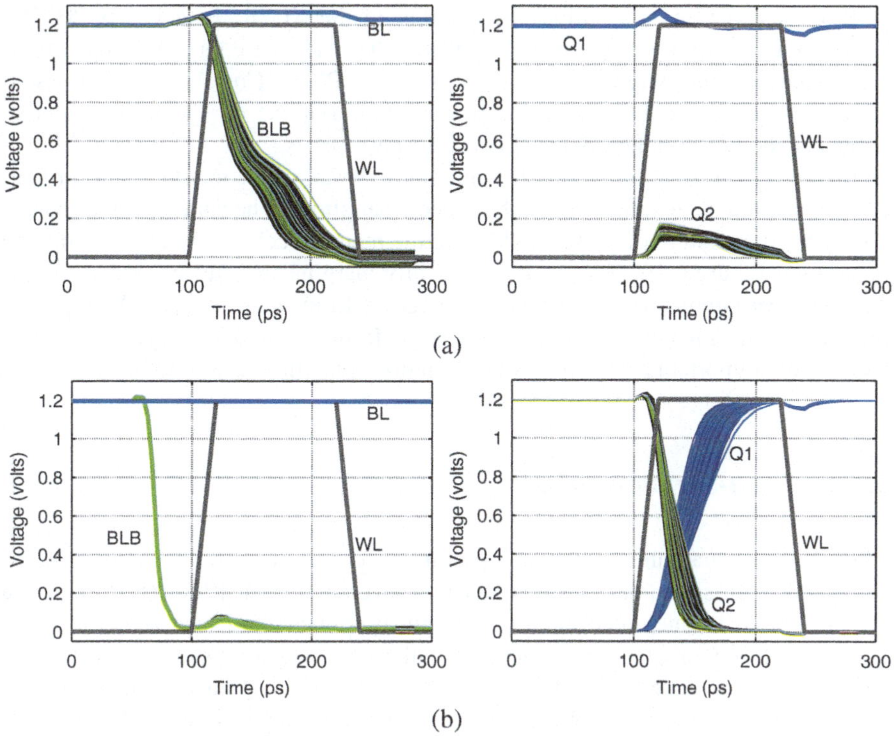

Fig. 6.18 Timing response of the SRAM cell under process parameter variations. (**a**) Reading operation under process variations. (**b**) Writing operation under process variations

Hint *The design of the memory cell must be carefully sized to guarantee correct reading, writing, and holding operations under process variations.*

6.4.3 Stability Margins Under Process Variations of the 6T-SRAM Cell

The stability margin is an important metric for the reliable operation of the memory cell. The static noise margin (SNM) provides the stability margin of the memory cell. The SNM gives the maximum voltage noise the memory cell can tolerate [3, 17, 23, 24]. The SNM for the memory cell is measured with a *butterfly* plot, which is obtained by plotting the voltage transfer characteristic of the inverter of the cell and its inverse. The SNM value is defined by the side of the largest square that can be drawn in the *butterfly* plot. Correct hold and read operations have two stable modes, and a correct write operation has only one stable mode.

Process variations may affect the stability of the 6T-SRAM cell [16, 20, 25]. The stability margins of the memory cell for the read, write, and hold operations are obtained using Monte Carlo analysis with SPICE. Correlated variations in parameters L, W, and T_{ox} and purely random variations of V_{th} are considered. Table 6.2 shows the standard deviation of the parameters.

Figure 6.19 illustrates the stability margins (SNMs) of the memory cell for both at nominal parameter values and under process variations. The value into the squares represents the amount of noise (SNM) that the memory cell can tolerate during the operations. Figure 6.19 shows that the stability margins are reduced under process parameter variations. The read margin decreases from 0.203 to 0.176 V, the write margin from 0.47 to 0.39 V, and the hold margin from 0.43 to 0.39 V. Previous results indicate the important role that process variations play in SRAM cell behavior.

6.4.4 High-Sigma Analysis

Chip designs guaranteeing correct performance for more than 3σ of process variations (high-sigma design) are a must on several products [26–28]. High-sigma design is used in integrated circuits for critical applications, such as aeronautics, automotive, and medical equipment. A failure in critical systems may lead to catastrophic consequences [19, 26, 29]. Also, the high-sigma design is used in components of integrated circuits that are replicated many times in large arrays [19, 26, 28, 29]. The 6T-SRAM memory cell is a component that usually needs to pass a high-σ analysis as this cell is replicated in large arrays. Furthermore, during the hold mode (data retention) operation, the bias voltage (VDD) of the memory cell is commonly lowered to reduce the leakage power consumption [19, 30]. Nevertheless, data stored in the cell can be destroyed when lowering VDD due to process variations.

Figure 6.20 illustrates the stability margin of the cell in the hold operation mode as function of VDD for both the nominal parameter values and under process variations (3σ and 6σ). The hold stability noise margin of the cell reduces as the power supply is lowered for all the analysis (nominal, 3σ and 6σ). The hold stability margin of the memory cell at high-sigma analysis, accounting for 6σ variations in the parameters, is lower than for conventional 3σ analysis. For the three analyzed cases (nominal, 3σ, and 6σ), the cell works well above 0.8 V of the supply voltage, but the memory cell may be at risk at lower supply voltages.

In Fig. 6.20, 1000 and 100,000 Monte Carlo run iterations in SPICE were considered for the analysis at 3σ and 6σ, respectively. 100,000 iterations to account 6σ is a reduced number. According to the theory, the number of iterations should be around one billion of iterations [26, 28], which is difficult to achieve with an electrical simulator as SPICE. Nevertheless, the analysis was made for illustrating purposes, and the results clearly show the tendency of the 6σ analysis. Solido design automation has developed a high-sigma Monte Carlo (HSMC) tool to make a high-sigma analysis lowering the number of simulations [28, 29].

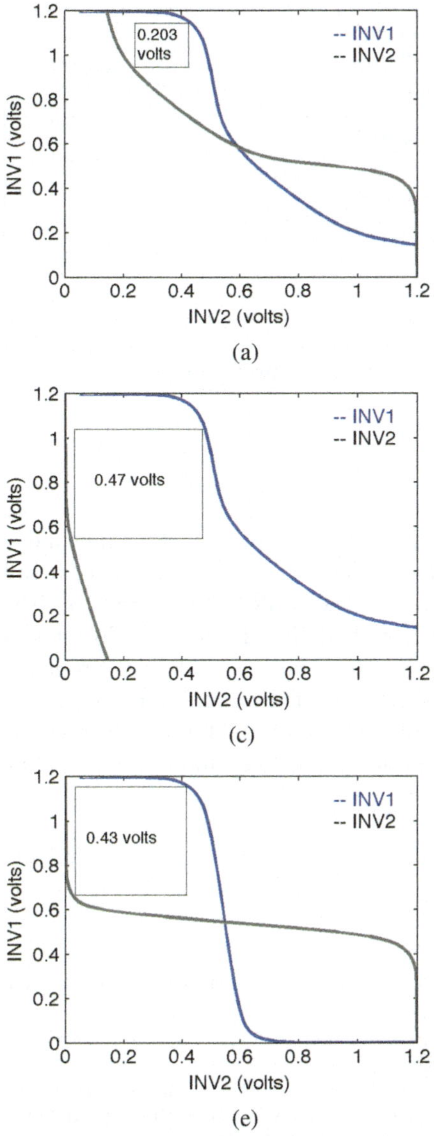

Fig. 6.19 Stability margins of the 6T-SRAM cell for the read, write, and hold operations at nominal and process variations. (**a**) Read margin at nominal. (**b**) Read margin with process variations. (**c**) Write margin at nominal. (**d**) Write margin with process variations. (**e**) Hold margin at nominal. (**f**) Hold margin with process variations

Hint *The design of a 6T-SRAM cell and other circuits that are replicated many times in an array must fulfill the high-sigma constraints to guarantee the SRAM reliability under process variations.*

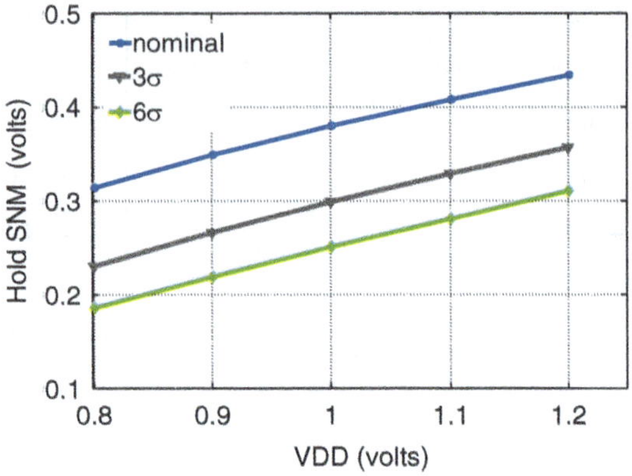

Fig. 6.20 6T-SRAM hold stability margin for both the nominal value of parameters and under process parameter variations (3σ and 6σ) as a function of the supply voltage value

6.5 Summary

The impact of process variations on the delay performance of some illustrative circuits of general interest is in this chapter studied. The circuits are a 1-bit full adder, a 4-bit RCA, the benchmark circuit S27, and a SRAM memory cell. The statistical delay of the circuits was obtained using both SPICE Monte Carlo analysis and the sensitivity delay model. Some important *hints* were derived from the analysis that may be used to improve performance of the circuits under process variations. The behavior of stability margins of the memory cell for both the nominal and under process parameter variations was analyzed. Even more, a high-sigma analysis for the memory cell was outlined.

References

1. N.H.E. Weste, D.M. Harris, *CMOS VLSI Design: A Circuits and Systems Perspective*, 4th edn. (Addison-Wesley, Reading, MA, 2010)
2. J.M. Rabaey, A. Chandrakasan, B. Nikolić, *Digital Integrated Circuits: A Design Perspective*, 2nd edn. (Prentice Hall, Englewood Cliffs, 2003)
3. H. Veendrick, *Nanometer CMOS ICs: From Basics to ASICs* (Springer, Cham, 2008)
4. K.M. Priyadarshini, M.N. Sabari, Comparative analysis of a low power and high speed hybrid 1-bit full adder for ULSI circuits. Int. J. Sci. Res. **5**(9), 1631–1635 (2016)
5. M. Suman, J. Samanta, D. Chowdhury, J. Bhaumik, Relative performance analysis of different CMOS full adder circuits. Int. J. Comput. Appl. **114**(6), 8–14 (2015)
6. M. Alioto, G. Palumbo, *Comparison on CMOS Full Adders in Different Design Styles with Emphasis on Low-Power Topologies*, University of Catania, Italy
7. SPICE2: a Computer Program to Simulate Semiconductor Devices, Memo ERL-M520, Department of Electrical Engineering and Computer Science, University of California at Berkeley (1975)
8. A. Thakur, D. Chilamakuri, D. Velenis, Effects of process and environmental variations on adder architectures, in *49th IEEE International Midwest Symposium on Circuits and Systems*, vol. 1 (2006), pp. 36–40. https://doi.org/10.1109/MWSCAS.2006.381988

9. D. Mohapatra, G. Karakonstantis, K. Roy, Low-power process-variation tolerant arithmetic units using input-based elastic clocking, in *ACM/IEEE International Symposium on Low Power Electronics and Design (ISLPED)* (2007). https://doi.org/10.1145/1283780.1283797
10. M. Lanuzza, F. Frustaci, S. Perri, P. Corsonello, Design of energy aware adder circuits considering random intra-die process variations. J. Low Power Electron. Appl. **1**(1), 97–108 (2011). https://doi.org/10.3390/jlpea1010097
11. N.A. Conos, S. Meguerdichian, M. Potkonjak, Gate sizing under uncertainty, in *VLSI-SoC: At the Crossroads of Emerging Trends. VLSI-SoC 2013*, ed. by A. Orailoglu, H. Ugurdag, L. Silveira, M. Margala, R. Reis. IFIP Advances in Information and Communication Technology, vol. 461 (Springer, Cham, 2015), pp. 23–47. https://doi.org/10.1007/978-3-319-23799-2_2
12. F. Brglez, D. Bryan, K. Kozminski, Combinational profiles of sequential benchmark circuits, in *IEEE International Symposium on Circuits and Systems ISCAS'89*, May 1989, vol. 3, pp. 1929–1934. https://doi.org/10.1109/ISCAS.1989.100747
13. M.M. Ozdal, S. Burns, J. Hu, Gate sizing and device technology selection algorithms for high-performance industrial designs, in *2011 IEEE/ACM International Conference on Computer-Aided Design (ICCAD)*, Nov 2011, pp. 724–731. https://doi.org/10.1109/ICCAD.2011.6105409
14. W. Chen, C.-T. Hsieh, M. Pedram, Simultaneous gate sizing and placement. IEEE Trans. Comput. Aided Des. Integr. Circuits Syst. **19**, 206–214 (2000). https://doi.org/10.1109/43.828549
15. Q.Y. Tang, P. Friedberg, G. Cheng, C.J. Spanos, Circuit size optimization with multiple sources of variation and position dependant correlation, in *Proceedings of SPIE – The International Society for Optical Engineering*, vol. 6521 (2007). https://doi.org/10.1117/12.711794
16. A. Pavlov, M. Sachdev, *CMOS SRAM Circuit Design and Parametric Test in Nano-scaled Technologies: Process-Aware SRAM Design and Test* (Springer, New York, 2008)
17. T.P. HAraszti, *CMOS Memory Circuits* (Springer, New York, 2000)
18. W.-K. Chen, *The VLSI Handbook* (CRC Press, Boca Raton, 2007)
19. C. Shin, *Variation-Aware Advanced CMOS Devices and SRAM* (Springer, Dordrecht, 2016)
20. A. Islam, M. Hasan, A technique to mitigate impact of process, voltage and temperature variations on design metrics of SRAM Cell. Microelectron. Reliab. **52**(2), 405–411 (2012). https://doi.org/10.1016/j.microrel.2011.09.034
21. E.I. Vatajelu, J. Figueras, Statistical analysis of 6T SRAM data retention voltage under process variation, in *IEEE 14th International Symposium on Design and Diagnostics of Electronic Circuits and Systems (DDECS)*, Apr 2011, pp. 365–370. https://doi.org/10.1109/DDECS.2011.5783112
22. M. Pelgrom, A. Duinmaijer, A. Welbers, Matching properties of MOS transistors. IEEE J. Solid-State Circuits **24**(5), 1433–1439 (1989). https://doi.org/10.1109/JSSC.1989.572629
23. E. Seevinck, F.J. List, J. Lohstroh, Static-noise margin analysis of MOS SRAM cells. IEEE J. Solid-State Circuits **22**(5), 748–754 (1987)
24. D. Mukherjee, H.K. Mondal, B.V.R. Reddy, Static noise margin analysis of SRAM cell for high speed application. IJCSI Int. J. Comput. Sci. Issues **7**(5), 175–180 (2010)
25. E. Grossar, M. Stucchi, K. Maex, W. Dehaene, Read stability and write-ability analysis of SRAM cells for nanometer technologies. IEEE J. Solid-State Circuits **41**(11), 2577–2588 (2006). https://doi.org/10.1109/JSSC.2006.883344
26. D. Doman, *Engineering the CMOS Library: Enhancing Digital Design Kits for Competitive Silicon* (Wiley, Hoboken, 2012)
27. Y. Cao, *Predictive Technology Model for Robust Nanoelectronic Design* (Springer, Boston, 2011)
28. Solido Design Automation Inc., *High-Sigma Monte Carlo for High Yield and Performance Memory Design* (2012)
29. T. McConaghy, K. Breen, J. Dyck, A. Gupta, *Variation-Aware Design of Custom Integrated Circuits: A Hands-on Field Guide* (Springer, New York, 2013)
30. S. Bhunia, S. Mukhopadhyay, *Low-Power Variation-Tolerant Design in Nanometer Silicon* (Springer, New York, 2011)

Chapter 7
Designing with FinFETs and Process Variation Impact

Abstract Scaled planar transistor devices present significant increases of the leakage current exacerbated by process variations, which limits the performance of some electronic applications. FinFET technology has been adopted starting 22 nm technology node for high-performance and power-efficient applications. This chapter introduces FinFET devices. The fabrication steps for making fins and middle-of-line (MOL) local interconnects are described. Design issues unique to FinFET technology are discussed. A step-by-step procedure to create the layout of an inverter cell is presented. The main sources of process variations in FinFET technology are analyzed, and their impact on the delay performance of logic cells is discussed. The computing of the delay variance (standard deviation) of an inverter gate based on FinFET technology is presented. A hand-by-hand example of computing the delay of a single logic cell is presented.

7.1 FinFET Devices

In a MOS planar transistor, the current channel is created into the substrate (Fig. 7.1a), but in 3D FinFET transistors, the current channel rises over the substrate (Fig. 7.1b). In FinFET devices, a metal gate wraps around a thin fin, which increases the gate control over the channel and reduces the short-channel effects [1]. This structure has current channels at the three walls of the fin. The current flows at the left and right sidewalls and also at the top (Fig. 7.1).

Fins can be built on SOI material or bulk silicon. SOI technology has the advantages of easy fabrication and scalability but presents some drawbacks such as higher defect density than in bulk, and the generated heat is difficult to dissipate due to the thick buried oxide [2]. On the other hand, Bulk FinFETs reduces the problems of SOI FinFETS [2].

V. Champac, J. G. Gervacio, *Timing Performance of Nanometer Digital Circuits Under Process Variations*, Frontiers in Electronic Testing 39,
https://doi.org/10.1007/978-3-319-75465-9_7

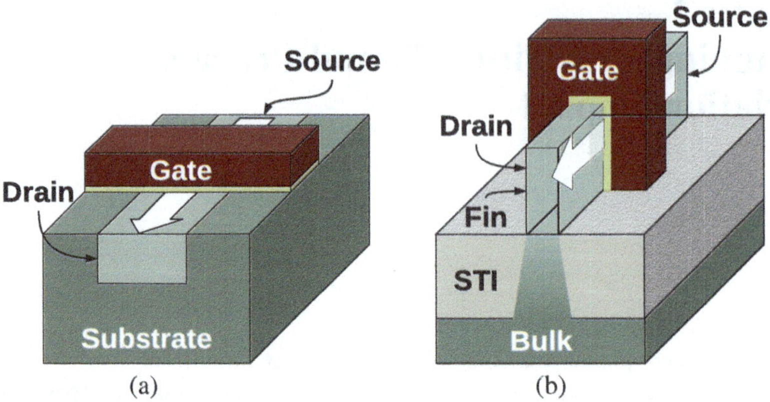

Fig. 7.1 FinFET device. (**a**) Planar transistor. (**b**) FinFET transistor

7.2 Making Fins and MOL Interconnects

7.2.1 Double Patterning

Traditional optical lithography techniques are facing resolution problems to transfer patterns to the wafer in the nanometer regime. The pattern resolution of a photolithographic process can be improved with a smaller wavelength and higher numerical aperture of the lens [3]. However, scaling of the wavelength stopped at 193 nm, and improving numerical aperture is not straightforward [3, 4]. New design and lithographic techniques have been adopted to overcome this technological limitation [3, 5].

Double patterning lithography is a technique to enhance resolution [6, 7]. In double patterning, the pitch size is doubled by decomposing the original pattern into two masks, which are processed separately. Figure 7.2 illustrates the use of double patterning. First, the pattern to the left is transferred to the wafer with one lithographic process, and then the pattern at the middle is transferred to the wafer with a second lithographic process. The final pattern transferred to the wafer is shown to the right.

7.2.2 Self-aligned Double Patterning (SADP)

Litho-etch-litho-etch (LELE) and self-aligned double patterning (SADP) are the two main lithographic process using double patterning [3]. Even more, SADP has become the accepted patterning process for 2× nm and further scaled technology nodes [8]. Figure 7.3 illustrates the creation of fins using SADP process. In this case, one litho-etch step followed by a spacer patterning takes place. First, the

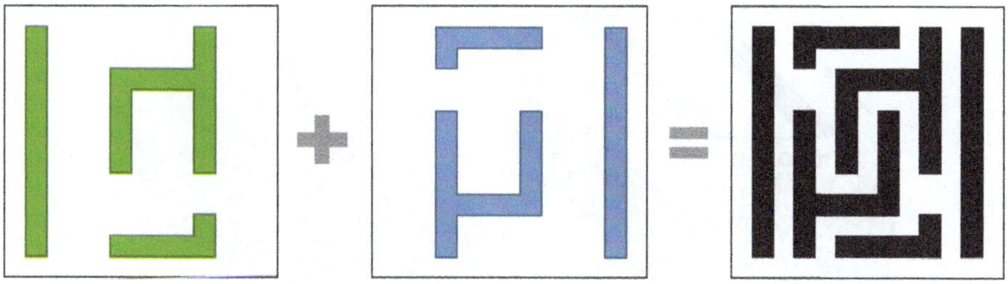

Fig. 7.2 Double patterning

PR is exposed to light according to the desired pattern defined in the photomask (Fig. 7.3a). Then, the PR is developed, and the hard mask is etched (Fig. 7.3b). A PR removal follows this. Second, a spacer layer is deposited (Fig. 7.3c), and then it is etched back. Only those portions of the spacer layer beside the walls of the hard mask, known as spacers, remain (Fig. 7.3d). The spacers have a desired width. The hard mask is removed. At this stage, the desired pattern has been transferred to the spacer layer (Fig. 7.3e). Then, the spacers are used as a mask for the silicon etch (Fig. 7.3f), and finally, the spacers are removed (Fig. 7.3g) [3]. As SADP lithography defines the fins using a spacer, they are referred as spacer-defined fins.

7.2.3 Middle-of-Line (MOL) Local Interconnects

In a traditional fabrication process of semiconductor ICs, the manufacturing process steps are grouped into two major stages, which are known as front end of line (FEOL) and the back end of line (BEOL) [9, 10]. Devices are created in FEOL, and all the connections between the devices are made in BEOL [9, 10]. Since 20 nm CMOS technology node, an intermediate interconnect schema known as middle-of-line (MOL) local interconnects was introduced between FEOL and BEOL stages [11, 12]. MOL allows connections of devices below BEOL. MOL connections are usually made of tungsten (W) [13], and the dielectric material between the gate and MOL is typically SiN [5]. Tungsten allows the filling of taller and tighter trenches but presents higher resistivity than copper (Cu).

The two main MOL layers are CA and CB [11, 14, 15]. CA layer is mostly used to connect drains (sources) of different fins and to connect the transistor to power/ground rails, and CB layer is used to connect multiple gates. Figure 7.4a shows the layout of an inverter composed of two fins and two fingers constructed using our in-house layout tool for FinFETs, and Fig. 7.4b shows the 3D structure of the designed inverter layout obtained using Raphael 3D solver (https://www.synopsys.com).

Let us first see how the source/drain electrodes of the transistors are connected to the first metal layer (M1) (Fig. 7.4) [11, 14, 15]. Before MOL, the source/drain regions are raised (RSD) in the fin outside the channel region (Fig. 7.4) to increase

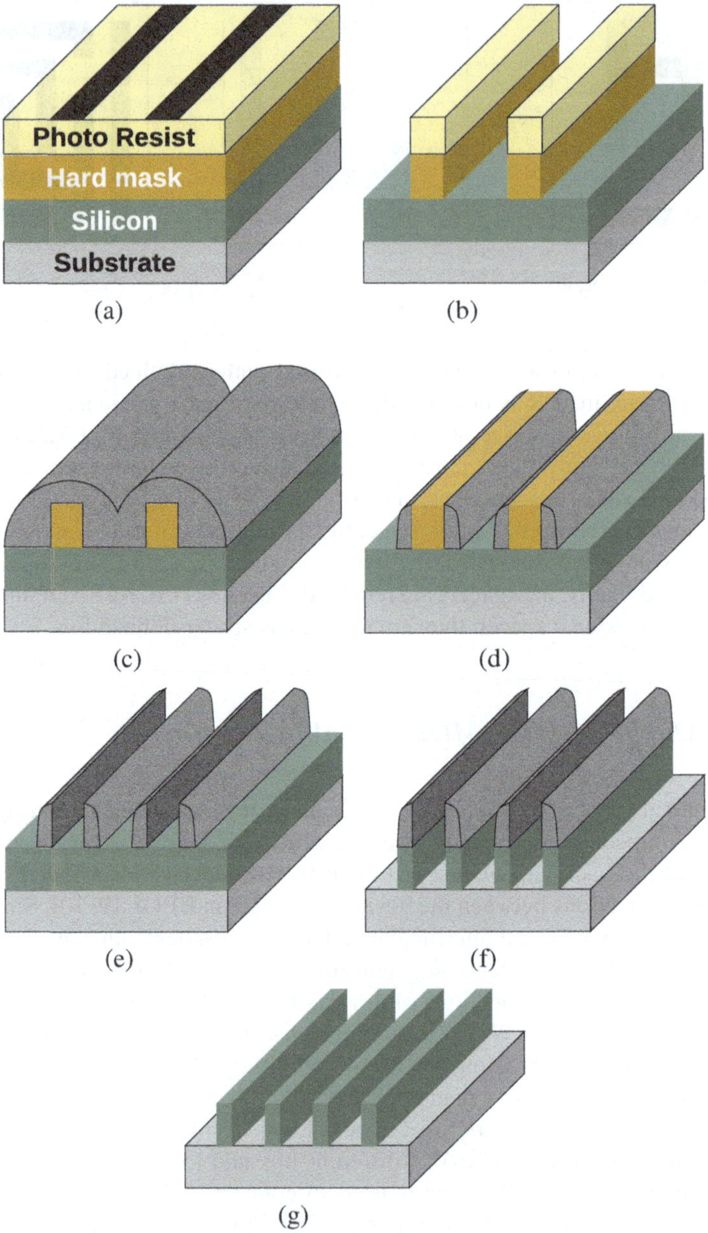

Fig. 7.3 Making fins with self-aligned double patterning (SADP) [3]. (**a**) PR exposure. (**b**) After PR development and hard mask etch. (**c**) Spacer deposition. (**d**) Spacer etch back. (**e**) Hard mask removal. (**f**) Silicon etch. (**g**) Spacer removal

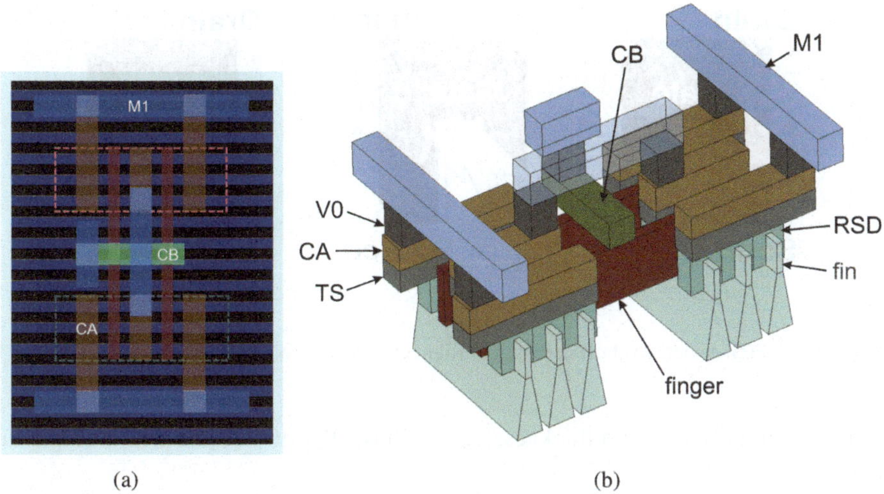

Fig. 7.4 Inverter using middle-of-line (MOL) local interconnects. (**a**) Layout of an inverter with two fins and two fingers using MOL. (**b**) 3D structure of the inverter layout

the cross section of these regions, which reduces their series resistance due to the thin fins [11]. A TS layer is placed over the RSD layer to make an electrical connection among the multiple fins of a single device. The top surface of the TS layer is at the same vertical level than the top surface of the gate. A CA layer is placed over the TS layer, and a V0 via is placed over CA layer to connect it to the first metal layer (M1) [16].

Now, let us see how the gate of the transistors are connected to M1 [11, 14, 15]. A CB layer is placed over the gates of different devices. The top surface of the CB contact is at the same vertical level than the top surface of the CA contact. A V0 via is placed over CB layer to connect it to the first metal layer (M1) [11].

Besides the increase of the connection resistance due to the use of tungsten instead of copper as MOL material, MOL presents other drawbacks. The parasitic capacitances increase due to small separation between TS/CA contacts and the gate, and a leakage current may exist between TS/CA and the gate [12]. However, the use of MOL allows the design of high-density standard cells [11]. At a 20 nm technology node, library cells constructed using MOL showed better power/performance, area, and cost in comparison to cells constructed without MOL [11].

7.3 Design with FinFETs

In 3D FinFETs, channel inversions are created in the three walls of the fin. Thus, the current flows at left, right, and top sides of the fin (Fig. 7.5). The effective channel width (W_{eff}) of a 3D FinFET depends on the geometrical parameters of the fin such

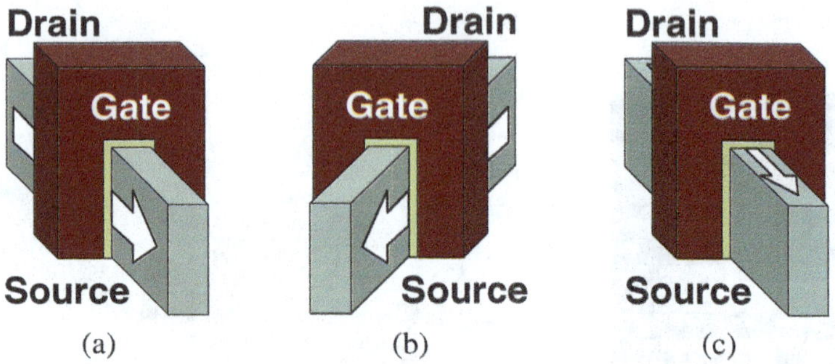

Fig. 7.5 FinFET device. (**a**) Left current. (**b**) Right current. (**c**) Top current

as fin height (H_{fin}) and fin thickness (T_{fin}). Thus, W_{eff} of a device with a single fin is given by

$$W_{\text{eff}} = 2H_{\text{fin}} + T_{\text{fin}} \tag{7.1}$$

As the manufacturing process defines T_{fin} and H_{fin}, designers make use of multiple fins and multiple fingers to be able to construct digital gates with different driving strengths [17, 18]. The designer can no longer have continuous values but discrete values of the effective channel width (W_{eff}), which now becomes a quantized parameter.

7.3.1 Multi-fin Device

FinFETs with multiple fins can be used by circuit designers to obtain the desired driving strength (Fig. 7.6a). The effective channel width of a multi-fin transistor depends on the number of fins (NFIN), and it is given by

$$W_{\text{eff}} = \text{NFIN} \cdot (2H_{\text{fin}} + T_{\text{fin}}) \tag{7.2}$$

NFIN = 6 for the example shown in Fig. 7.6a.

7.3.2 Multi-finger Device

The fins can be crossed by several gates (known as fingers), and then the fingers are connected using CB layer to increase the driving strength of the device further

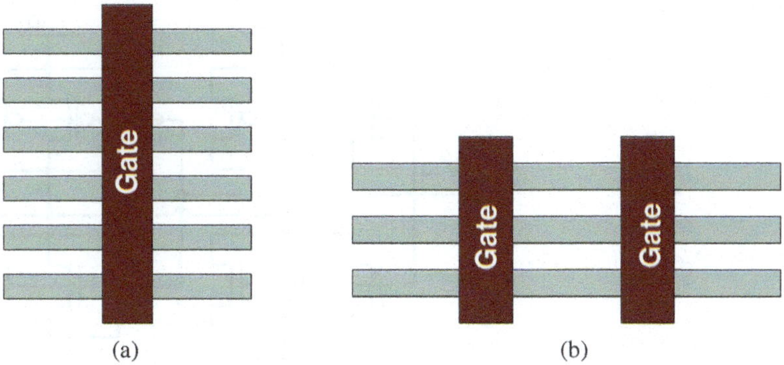

Fig. 7.6 Multi-fin and multi-fin multi-finger devices. (**a**) Multi-fin device, NFIN = 6 NF = 1. (**b**) Multi-fin multi-finger device, NFIN = 3 NF = 2

(Fig. 7.6b). The effective channel width of a multi-fin multi-finger transistor depends on the number of fins (NFIN) and fingers (NF), and it is given by

$$W_{\text{eff}} = \text{NFIN} \cdot \text{NF} \cdot (2H_{\text{fin}} + T_{\text{fin}}) \tag{7.3}$$

NFIN = 3 and NF = 2 for the example shown in Fig. 7.6b.

7.4 Layout Design with FinFETs

Some layout strategies of logic cells based on FinFET technology appear in the literature [14, 19].

Figure 7.7 shows the transistor level schematic of an inverter cell composed of three fins and two fingers, and Fig. 7.8 illustrates a step-by-step procedure to design its layout. The layout description is based on the literature, but some differences may appear in industrial layouts. The procedure starts with a sea of horizontal fins which are equidistant. The first step consists of drawing the n-well region, as well as the n-type and p-type implants, to define the nMOS and pMOS devices (Fig. 7.8a). It must be noted that each implant covers three fins. For the next step, the gates are defined (Fig. 7.8b), the two inner fingers are used as active gates, and the two outer fingers are used as dummy gates for process uniformity. Multiple fins are connected using CA layer to provide access to the regions of drain/source of the transistors (Fig. 7.8c).

The two inner fingers are connected using CB layer (Fig. 7.8c). This connection corresponds to the input of the inverter. V0 vias are placed on top of CA and CB layers (Fig. 7.8d). Metal layer M1 is used to connect the V0 vias of interest. Metal M1 connects vias V0 to obtain the input and output of the inverter and also to have power and ground rails (Fig. 7.8e). Cut masks are placed in regions to be removed (Fig. 7.8f). Figure 7.8g shows the final layout.

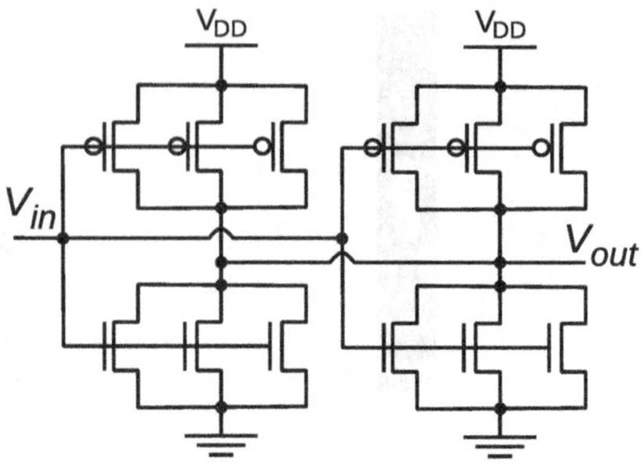

Fig. 7.7 Schematic of FinFET inverter with a three fins and two fingers

7.5 Sources of Process Variations in FinFET Technology

This section discusses the gate work function variations (WFVs) and line edge roughness (LER), which are two important sources of variations in FinFET technology [20]. The impact of random dopant fluctuations (RDF) that reduces in a FinFET device due to the use of non-doped or barely doped channel is studied in [21]. Fin height variations and other sources of variations are studied in [22].

7.5.1 Work Function Variation (WFV)

The transistor threshold voltage depends on the flat band voltage (V_{FB}), which is determined by the difference between the gate work function (Φ_M) and the semiconductor work function (Φ_S) [23]. Thus, $V_{\text{FB}} = \Phi_M - \Phi_S$. At 65 nm technology, the insulating dielectric composed of a few atoms was losing its insulating properties as it turned leaky. Because of this, a thicker high-*k* dielectric was introduced at 45 nm technology node to counteract the leakage current [24, 25]. As the high-*k* dielectric materials do not work fine with polysilicon gates, the material of the transistor gate was changed to metal [25], whose work function presents random variations from one device to another device. The cause of the random variations is that the work function of a metal depends on its grain orientations [26], which are stochastically determined during the growth period [27]. Variability of the metal gate granularity has been found an important source of variability with the introduction of high-*k* metal gates [26–28]. Figure 7.9 illustrates a FinFET device with a magnification of the metal gate, which may have different grain orientations. Grains in the metal gate may have different work function with different occurrence probabilities. Because of the random nature of the grain orientation, the work function of a metal gate (Φ_M) is represented by a random variable with an expected value and variance [27].

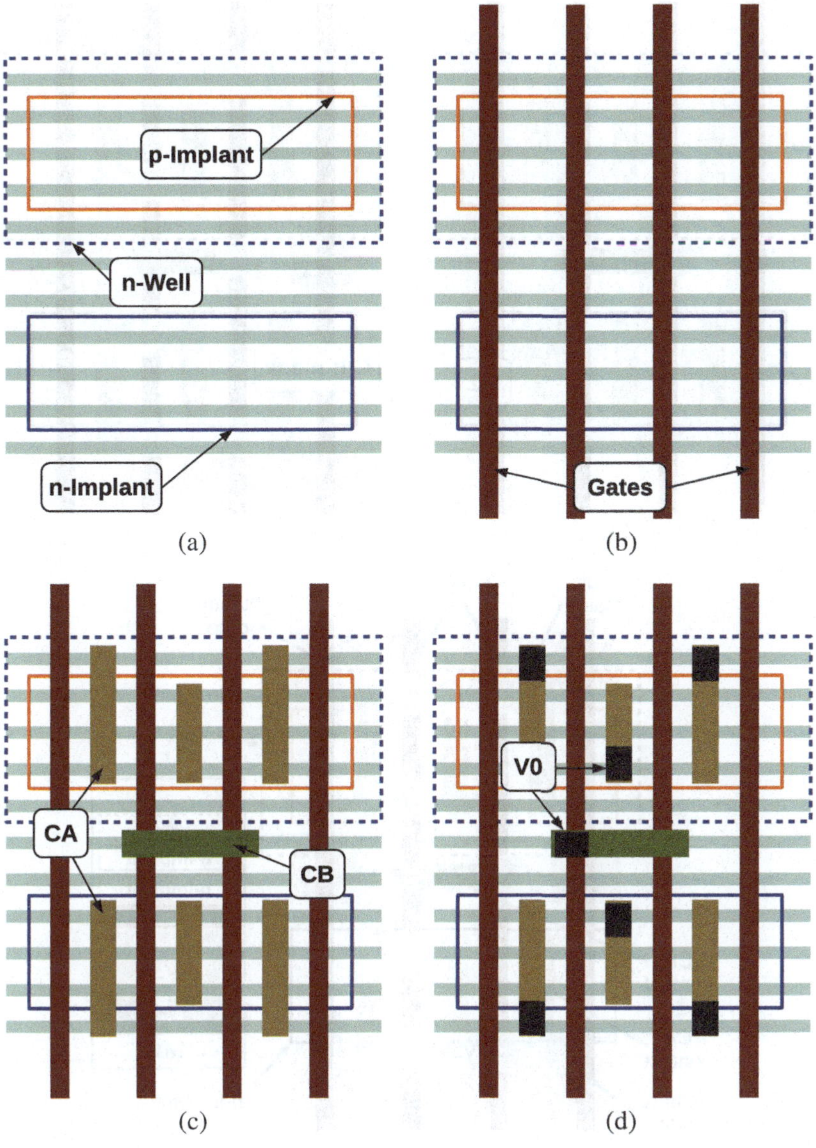

Fig. 7.8 FinFET layout design. (**a**) Sea of fins with n-well and implants. (**b**) Gates placement. (**c**) CA and CB contacts. (**d**) V0 vias. (**e**) M1 connections. (**f**) Cut mask. (**g**) V0 vias

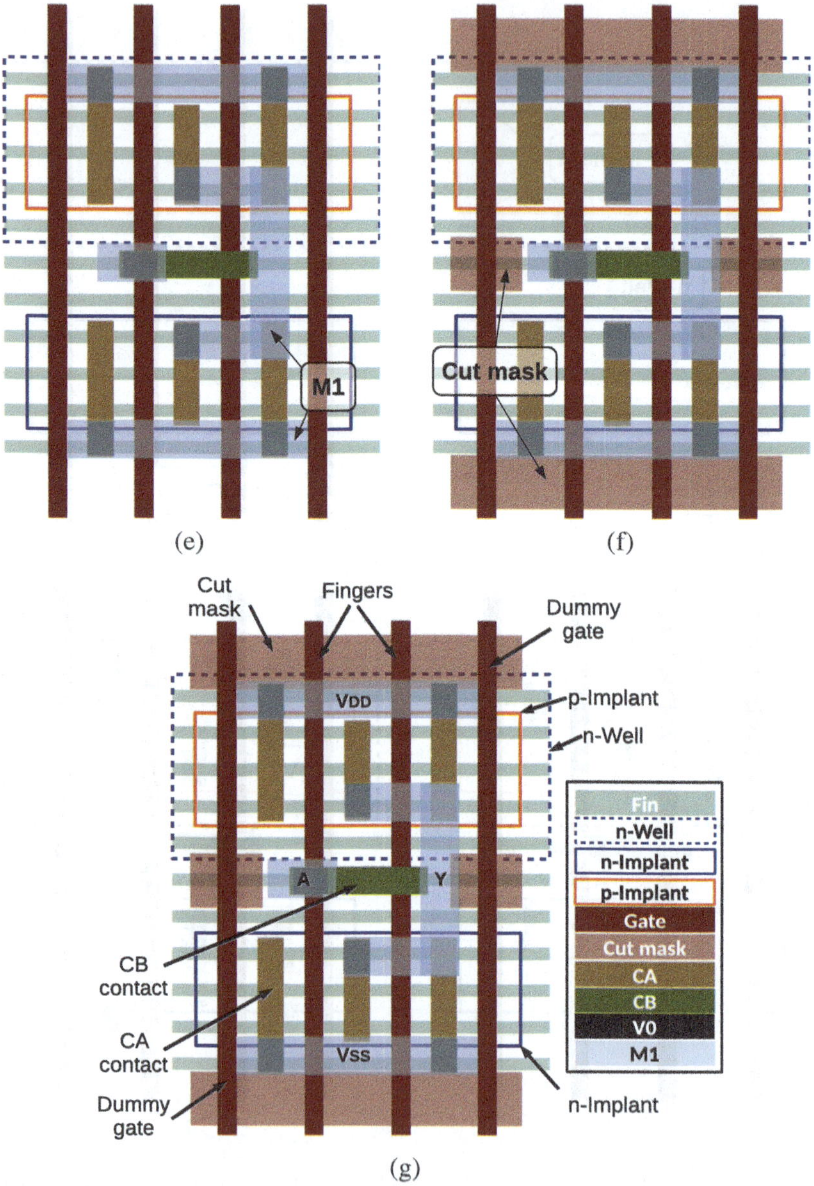

Fig. 7.8 (continued)

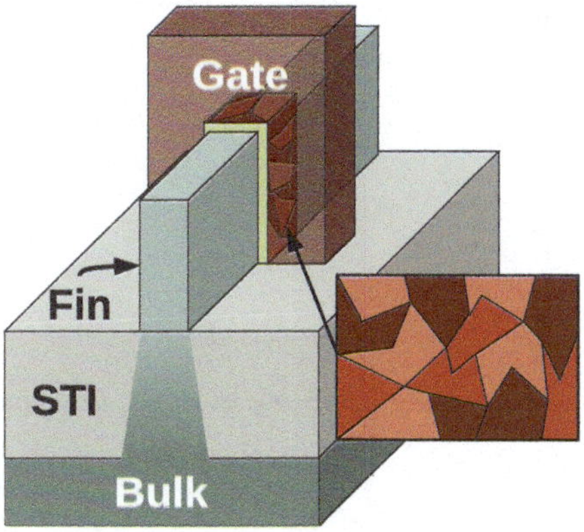

Fig. 7.9 Schematic of metal gate with different grain orientations [27]

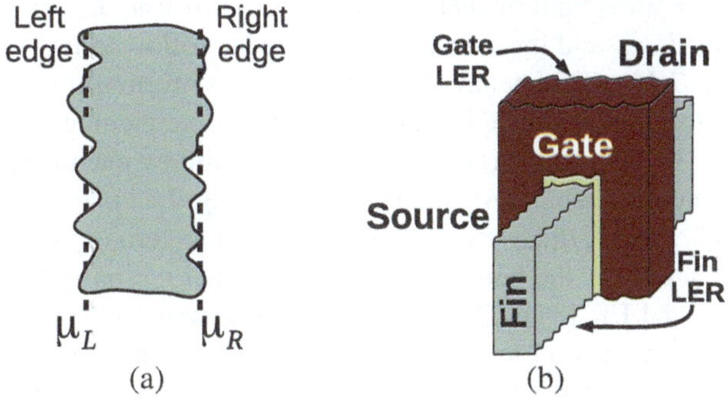

Fig. 7.10 FinFET LER. (**a**) Line Feature LER. (**b**) FinFET LER

7.5.2 Line Edge Roughness (LER)

Line edge roughness (LER) refers to random variations of a line edges up and down along their mean value (Fig. 7.10a). They are caused by stochastic behavior in some of the steps to transfer a pattern drawn in a mask to the desired layer. Among the causes, it can be mentioned variations on the exposure to light, statistical distributions of chemical species in the resist, and variations in chemical amplification [29, 30]. For advanced technologies, LER variations have become an important source of variability for feature sizes of sub-45 nm [28, 31]. They may impact FinFET transistor behavior by variations in the fin and metal gate [32].

The distance from one edge to another edge of a feature line (Fig. 7.10a) defines the line width roughness (LWR), which is the width of the feature line. The variations of LWR are expressed by Patel et al. [33]

$$\sigma_{\mathrm{LWR}}^2 = \sigma_L^2 + \sigma_R^2 - 2\rho_X\sigma_L\sigma_R \tag{7.4}$$

where σ_L and σ_R are the left and right edge standard deviations, and ρ_X is the cross-correlation coefficient between the two edges [33].

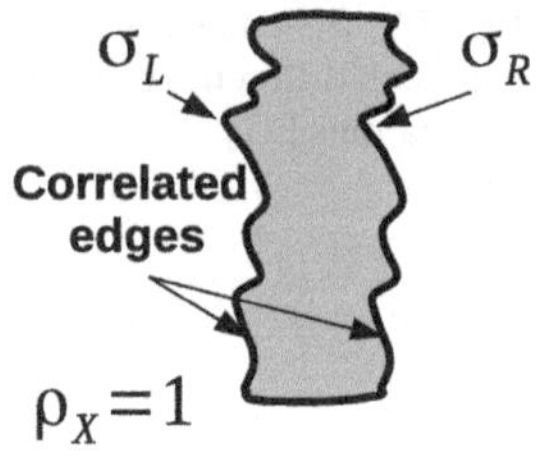

Fig. 7.11 Spacer-defined edges of a feature line

Assuming $\sigma_L = \sigma_R = \sigma_{\mathrm{LER}}$, (7.4) can be simplified to [33],

$$\sigma_{\mathrm{LWR}}^2 = 2\sigma_{\mathrm{LER}}^2(1 - \rho_X) \tag{7.5}$$

For a spacer-defined pattern such as with SADP process, the fins are defined by the spacers (Fig. 7.3f), and the variations in one edge follow to the variations in the other edge (Fig. 7.11). This behavior is due to the conformal thin-film deposition process that is followed by a highly uniform anisotropic etching during the spacer fabrication [30, 33]. As the variations at the edges follow each other, the random variations at the edges can be assumed correlated (Fig. 7.11), and they cancel each other (Eq. (7.4)). Hence, σ_{LWR} can be assumed equal to zero ($\sigma_{\mathrm{LWR}} = 0$).

Even that $\sigma_{\mathrm{LWR}} = 0$, the body shape in the channel region suffers bending due to the presence of LER [34]. This curving of the channel is subject to a random process. Thus, these variations may be considered as random variations in the effective channel length of the device [34]. Also, achieving perfectly correlated edges is a difficult task for a lithography process. The spacer-defined pattern may have correlations lower than one due to variations in the deposition and etching process or annealing steps [34]. Because of this, some small value of σ_{LWR} may remain.

LER in the fins causes fluctuations in the fin thickness, T_{fin}, (Fig. 7.10b), and it is named as Fin LER and also as FER. LER in the fins affects the effective channel width, the threshold voltage due to quantum mechanical confinement, and the generation-recombination component of the drain current [32].

LER in the gate causes fluctuations in the gate length, L_g, (Fig. 7.10b), and it is named as Gate LER and also as GER. LER in the gate affects the electron mobility, the threshold voltage due to short-channel effects, and the generation-recombination component of the drain current [32].

7.6 Computing the Delay Variance of a Logic Cell

This section studies the impact of process variations on the timing performance of logic cells based in FinFETs [34–37]. The delay variance of a multi-fin FinFET inverter built with a spacer-defined lithography process (SADP) is studied. The work function variation (WFV) and line edge roughness variations affecting the channel length (GER) and fin thickness (FER) are considered. In this study, the statistical performance of an inverter is illustrated for parameters having pure random or inter-die variations.

7.6.1 General Delay Expressions for Pure Random and Inter-die Variations

In Eq. (3.4), the process parameter models inter-die, intra-die correlated and intra-die pure variations. Our analysis of the delay performance of a multi-fin FinFET inverter considers only the impact of inter-die variations (also known as global variations) and pure random variations. The inclusion of spatially correlated parameter can be easily accounted as shown in Chap. 4. First, an expression to compute the standard deviation of the delay of a multi-fin inverter with a parameter subject only to pure random variations is developed. Then, a similar expression is obtained for a parameter subject only to inter-die variations.

7.6.1.1 Pure Random Variations

Let us focus on the high-to-low propagation delay of an inverter. A parameter subject to pure random variations in the transistor defined by n parallel fins is considered (Fig. 7.12). Pure random variations are represented by random variables that affect each fin independently. As pure random variations are independent variations on each fin, Eq. (A.11) of Appendix A is used to compute the delay standard deviation of a multi-fin inverter for pure random variations as follows:

$$\sigma_D^2 = (S_{D,p} \cdot \sigma_p)_1^2 + (S_{D,p} \cdot \sigma_p)_2^2 + \cdots + (S_{D,p} \cdot \sigma_p)_n^2 \tag{7.6}$$

where $S_{D,p}$ is the delay sensitivity to parameter p and σ_p is the standard deviation of the parameter p.

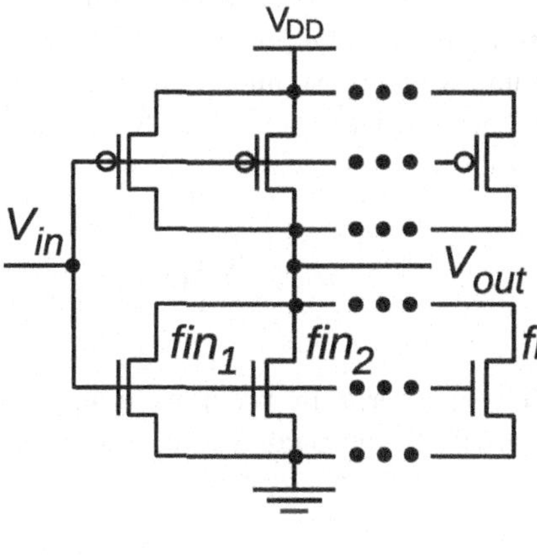

Fig. 7.12 Schematic of a multi-fin inverter with N fins

The delay sensitivity to a parameter is obtained by making variations of parameter p only in one fin, while the other parameters are held at their nominal values at all the fins. Thus, the delay sensitivities of each fin in Eq. (7.6) are the same. Also, all the standard deviations in expression (7.6) are the same as we are referring to the same process parameter. Based on the previous observations and replacing NFIN instead of n fins, a general expression to compute the variance of the delay due to pure random variations of a parameter p in a multi-inverter is as follows:

$$\sigma_D^2 = \text{NFIN} \cdot (S_{D,p} \cdot \sigma_p)^2 \tag{7.7}$$

where NFIN is the number of parallel fins, $S_{D,p}$ is the delay sensitivity to parameter p, and σ_p is the standard deviation of the process parameter.

7.6.2 *Inter-die or Global Variations*

Let us focus on the high-to-low propagation delay of the inverter. A parameter subject to inter-die variations in the transistors defined by n parallel fins is considered. As inter-die variations affect all the fins in the cell in the same way, Eq. (A.10) of Appendix A is used to compute the delay standard deviation of a multi-fin inverter for inter-die variations as follows:

$$\sigma_D^2 = \sum_{i=1}^{n} (S_{D,p}\sigma_p)_i^2 + 2\sum_{i=1}^{n}\sum_{j=i+1}^{n} (S_{D,p}\sigma_p)_i \cdot (S_{D,p}\sigma_p)_j \cdot \rho_{i,j} \tag{7.8}$$

where $S_{D,p}$ is the delay sensitivity, σ_p is the standard deviation of the parameter, and $\rho_{i,j}$ is the correlation of the parameter between a pair of fins.

The delay sensitivity to a parameter is obtained by making variations of parameter p only in one fin, while the other parameters are held at their nominal values at all the fins. Thus, the delay sensitivities of each fin in Eq. (7.8) are the same. Also, all the standard deviations in expression (7.8) are the same as we are referring to the same process parameter. As inter-die variations affect all the fins in the same way, the parameter variations between fins in a logic cell can be assumed fully correlated ($\rho_{i,j} = 1$) [38]. Based on the previous observations, Eq. (7.8) simplifies to

$$\sigma_D^2 = \sum_{i=1}^{n} (S_{D,p}\sigma_p)_i^2 + 2\sum_{i=1}^{n}\sum_{j=i+1}^{n} (S_{D,p}\sigma_p)^2 \tag{7.9}$$

The first term in previous equations repeats n times the squared product $(S_{D,p}\sigma_p)^2$. In the second term, the number of terms of the squared products can be obtained by computing the number of combinations of two elements in a set of n elements without repetition. The number of terms is

$$C_{n,k} = \frac{n!}{k!(n-k)!} \tag{7.10}$$

where n is the total of elements and k is the set of combinations.

Using $k = 2$ and after some operations, the previous expression reduces to

$$C_{n,k} = \frac{1}{2}(n^2 - n) \tag{7.11}$$

Replacing Eq. (7.11) and NFIN instead of n fins in Eq. (7.9), a general expression to compute the variance of the delay due to inter-die variations of a parameter p in a multi-fin inverter results as follows:

$$\sigma_D^2 = \text{NFIN}^2 \cdot (S_{D,p} \cdot \sigma_p)^2 \tag{7.12}$$

where NFIN is the number of parallel fins, $S_{D,p}$ is the delay sensitivity to parameter p, and σ_p is the standard deviation of the parameter.

Equations (7.7) and (7.12) are also valid for the low-to-high propagation delay of a multi-fin inverter as the pMOS network is also composed of n parallel fins.

7.6.3 Computing the Delay Variance of a Multi-fin Inverter

The delay performance of a multi-fin inverter that uses spacer-defined lithography process (SADP) is analyzed. Pure random variations in the work function variation (Φ_M) and inter-die variations in the fin thickness (T_{fin}) and channel length (L_g) are considered. Local variations for T_{fin} (FER) and L_g (GER) are assumed less significant. These variations may exist due to non-idealities in the process as previously discussed, and their inclusion is straightforward.

Using Eqs. (7.7) and (7.12), an expression to compute the variance of the high-to-low propagation delay time as a function of the number of fins of the multi-fin spacer-defined inverter is as follows:

$$\begin{aligned} \sigma_D^2 = {} & \text{NFIN}\left[(S_{D,\Phi_M}\sigma_{\Phi_M})^2\right] \\ & + \text{NFIN}^2\left[(S_{D,T_{\text{fin}}}\sigma_{T_{\text{fin}}})^2 + (S_{D,L_g}\sigma_{L_g})^2\right] \end{aligned} \tag{7.13}$$

where $\boldsymbol{\sigma}_{\boldsymbol{\Phi}_M}$ is the standard deviation of the parameter $\boldsymbol{\Phi}_M$, $\boldsymbol{\sigma}_{\boldsymbol{T}_{\text{fin}}}$ and $\boldsymbol{\sigma}_{\boldsymbol{L}_g}$ are the standard deviation of $\boldsymbol{T}_{\text{fin}}$ and L_g, respectively, and $\boldsymbol{S}_{\boldsymbol{D}}$, $\boldsymbol{\Phi}_M$, $\boldsymbol{S}_{\boldsymbol{D}}$, $\boldsymbol{T}_{\text{fin}}$, and $\boldsymbol{S}_{\boldsymbol{D}}$, $\boldsymbol{L}_g$ are the delay sensitivities in the nMOS network to small changes in $\boldsymbol{\Phi}_M$, $\boldsymbol{T}_{\text{fin}}$, and $\boldsymbol{L}_g$ respectively.

The delay sensitivities are estimated as stated previously.

Table 7.1 Standard deviations of the process parameters [33, 35, 43–45]

Parameter	Variation
Φ_M	$3\sigma_{\Phi_M} = 90\,\text{mV}$
T_{fin}	$3\sigma_{T_{\text{fin}}} = 0.8\,\text{nm}$
L_g	$3\sigma_{L_g} = 1.4\,\text{nm}$

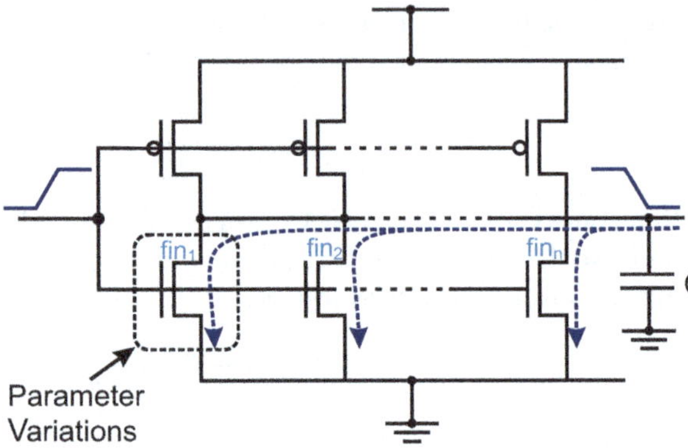

Fig. 7.13 Delay sensitivity methodology for multi-fin inverter

7.6.4 *Example of Computing the Delay Variance of a Multi-fin Inverter*

Multi-fin inverters with a different number of fins are built based on Intel's 14 nm FinFET technology [39]. Raphael 3D solver (https://www.synopsys.com) is used to extract the fringing capacitances of the structure. Predictive 14 nm PTM-MG bulk model [40] with SPICE BSIM-CMG standard FinFET compact model [41, 42] is used. Table 7.1 shows the standard deviations of the process parameters [35, 43, 44].

A step-by-step example to compute the delay variance of multi-fin inverters with different number of n fins (Fig. 7.12) is illustrated. The logic cells have a fan-out of four inverters with three fins for both the nMOS and pMOS transistors. Next, the delay sensitivities are estimated with SPICE simulations for the high-to-low and low-to-high propagation delays.

7.6.4.1 Estimation of the Delay Sensitivities

The delay sensitivity to a parameter is obtained by plotting the high-to-low (or low-to-high) propagation delay against the parameter value. The parameter is varied around its nominal value. Then, a linear fit of the curve is obtained, and the slope of this curve corresponds to the delay sensitivity. For the multi-fin inverter with n fins in both the nMOS and pMOS networks, the delay sensitivity to a parameter is obtained by making variations of a single parameter only in one of the fins of the switching network, while the other parameters are held at their nominal values at all the fins. Figure 7.13 illustrates this methodology for the high-to-low propagation

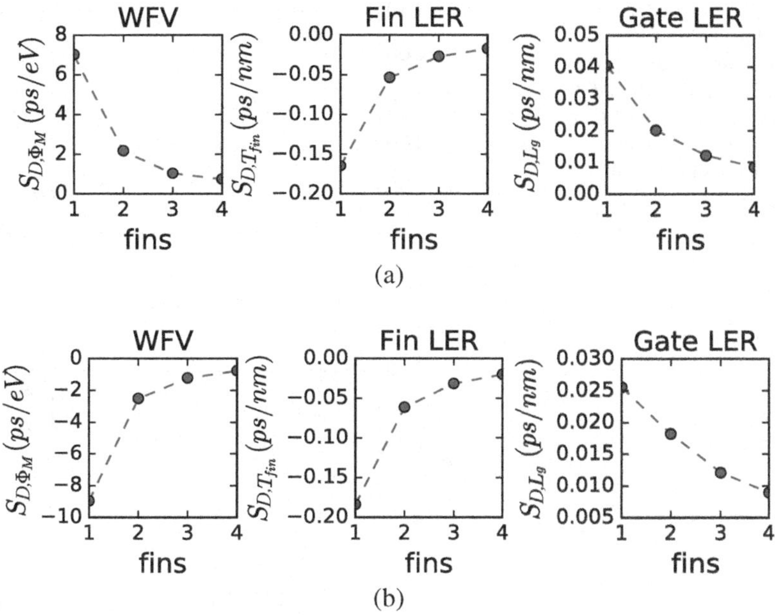

Fig. 7.14 Delay sensitivities for inverters with different number of fins. (**a**) High-to-low propagation delay time. (**b**) Low-to-high propagation delay time

delay. For the $0 \rightarrow 1$ input transition, the output discharges through all the n fins of the nMOS network. A single parameter (Φ_M or T_{fin} or L_g) is varied for only one of the fins in the nMOS network.

Figure 7.14 shows the obtained delay sensitivities by SPICE simulations. The delay sensitivities are inversely proportional to the number of fins. The impact of the variations in a single fin on the delay reduces as the number of fins increases.

7.6.4.2 Computing the Delay Variance

Using the delay sensitivities (Fig. 7.14) and the standard deviations of the process parameters (Table 7.1), the delay variance of a multi-fin inverter is computed with Eq. (7.13). The delay standard deviation is obtained by the square root of the variance. Figure 7.15 plots the contribution of each process parameter to the standard deviation of the delay for multi-fin inverters with a different number of fins. The solid traces represent the data obtained with SPICE, and the markers represent the data obtained with Eq. (7.13). The contribution of each process parameter to the standard deviation of the delay decreases as the number of fins increases. Φ_M has the most significant impact, followed by T_{fin} and L_g. Figure 7.16 shows the total standard deviation of the high-to-low and low-to-high propagation delays. Only the value of the standard deviation of the gate delay computed using Eq. (7.13) is shown.

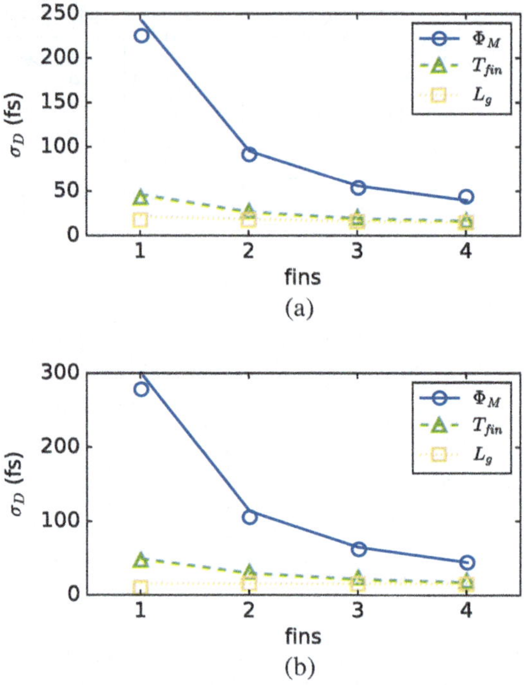

Fig. 7.15 Inverter delay standard deviation (spacer-defined); lines are for the SPICE simulation, and markers are for the analytical model. (**a**) Falling output (t_{PHL}). (**b**) Rising output (t_{PLH})

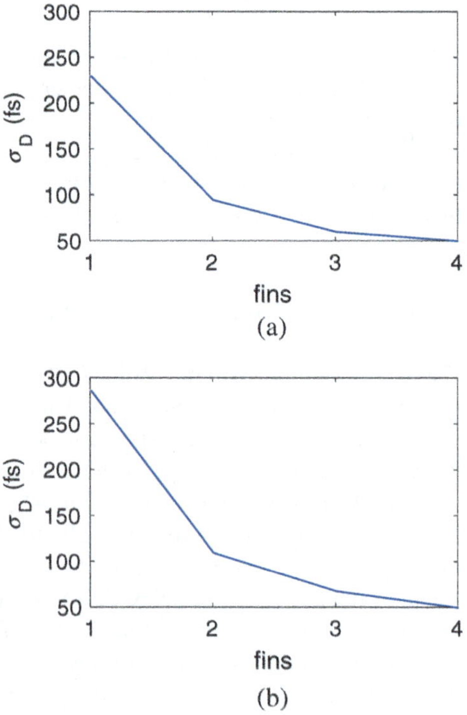

Fig. 7.16 Total standard deviation for multi-fin inverters with different number of fins. (**a**) Falling output (t_{PHL}). (**b**) Rising output (t_{PLH})

7.7 Summary

In this chapter, some important technology challenges related to circuits implemented in FinFET technology were studied. Double patterning, self-aligned double patterning (SADP) process for making fins, and middle-of-line (MOL) local interconnects were described. Key design issues to make successful circuits in this technology were presented such as multi-fin and multi-finger techniques. A step-by-step procedure to create the layout of an inverter cell was presented. Two important sources of process variations in FinFET technology were analyzed such as the work function variations (WFVs) and line edge roughness (LER). The impact of process variations on the performance of the delay variance of an inverter gate is discussed. The computing of the delay variance of a multi-fin inverter gate based on FinFET technology was presented. A hand-by-hand example of computing the delay of a multi-fin inverter gate was presented.

References

1. B. Ho, X. Sun, C. Shin, T.-J.K. Liu, Design optimization of multigate bulk MOSFETs. IEEE Trans. Electron Devices **60**(1), 28–33 (2013)
2. J.-H. Lee, *Nano Devices and Circuit Techniques for Low-Energy Applications and Energy Harvesting* (Springer, Dordrecht, 2016)
3. C.A. Mack, Seeing double. IEEE Spectrum **11**(45), 46–51 (2008). https://doi.org/10.1109/MSPEC.2008.4659384
4. C.C. Chiang, J. Kawa, *Design for Manufacturability and Yield for Nano-Scale CMOS* (Springer, Dordrecht, 2007)
5. M. Amirtharaj, Z. Kruder, Lecture: ENEE 416: Integrated Circuit (IC) Fabrication Lab (Fall 2011): Topic: Double patterning and hyper-numerical aperture immersion lithography (2011). http://www.ece.umd.edu/class/enee416/GroupActivities/GroupActivity8.htm
6. G. Bailey, A. Tritchkov, J. Park, L. Hong, V. Wiaux, E. Hendrickx, S. Verhaegen, P. Xie, J. Versluijs, Double pattern EDA solutions for 32nm HP and beyond, in *Proceeding of SPIE*, vol. 6521 (2007). https://doi.org/10.1117/12.712773
7. J. Park, S. Hsu, D. Van Den Broeke, J. Chen, M. Dusa, R. Socha, J. Finders, B. Vleeming, A. van Oosten, P. Nikolsky et al., Application challenges with double patterning technology (DPT) beyond 45nm, in *Proceeding of SPIE*, vol. 6349 (2006). https://doi.org/10.1117/12.692921
8. M.-S. Kim et al., Self-aligned double patterning of Ix nm FinFET; new device integration through the challenging geometries, in *2013 14th International Conference on Ultimate Integration on Silicon (ULIS)*, Mar 2013, pp. 101–104. https://doi.org/10.1109/ULIS.2013.6523501
9. M. Orshansky, S.R. Nassif, D. Boning, *Design for Manufacturability and Statistical Design: A Constructive Approach* (Springer, Boston, 2008)
10. G. Bouche, E. Geiss, S. Beasor, A. Wei, D.E. Civay, *Methods for Fabricating Finfet Integrated Circuits with Simultaneous Formation of Local Contact Openings*, Google Patents, US Patent App. 14/164,582 (2014)
11. M. Rashed, N. Jain, J. Kim, M. Tarabbia, I. Rahim, S. Ahmed, J. Kim, I. Lin, S. Chan, H. Yoshida, et al., Innovations in special constructs for standard cell libraries in sub 28nm technologies, in *IEEE International Electron Devices Meeting*, Dec 2013, pp. 9.7.1–9.7.4. https://doi.org/10.1109/IEDM.2013.6724597

12. T. Kauerauf, A. Branka, G. Sorrentino, P. Roussel, S. Demuynck, K. Croes, K. Mercha, J. Bömmels, Z. Tőkei, G. Groeseneken, Reliability of MOL local interconnects, in *IEEE International Reliability Physics Symposium (IRPS)*, Apr 2013, pp. 2F.5.1–2F.5.5. https://doi.org/10.1109/IRPS.2013.6531970
13. P. Schuddinck, M. Badaroglu, M. Stucchi, S. Demuynck, A. Hikavyy, M. Garcia-Bardon, A. Mercha, A. Mallik, T. Chiarella, S. Kubicek, et al., Standard cell level parasitics assessment in 20nm BPL and 14nm BFF, in *IEEE International Electron Devices Meeting (IEDM)*, Dec 2012, pp. 25.3.1–25.3.4. https://doi.org/10.1109/IEDM.2012.6479101
14. K. Vaidyanathan, Q. Zhu, L. Liebmann, K. Lai, S. Wu, R. Liu, Y. Liu, A. Strojwas, L. Pileggi, Exploiting sub-20-nm complementary metal-oxide semiconductor technology challenges to design affordable systems-on-chip. J. Micro/Nanolithogr. MEMS MOEMS **14**(1), 011007-1–011007-17 (2015). https://doi.org/10.1117/1.JMM.14.1.011007
15. R. Xie et al., A 7nm FinFET technology featuring EUV patterning and dual strained high mobility channels, in *IEEE International Electron Devices Meeting (IEDM)*, Dec 2016, pp. 2.7.1–2.7.4. https://doi.org/10.1109/IEDM.2016.7838334
16. Y. Lin, B. Yu, B. Xu, D.Z. Pan, Triple patterning aware detailed placement toward zero cross-row middle-of-line conflict, in *Proceedings of the IEEE/ACM International Conference on Computer-Aided Design*, vol. 36, Issue 7, pp. 1140–1152 (2015). https://doi.org/10.1109/TCAD.2017.2648843
17. T. An, K. Choe, K.-W. Kwon, S. Kim, Performance optimization study of FinFETs considering parasitic capacitance and resistance. J. Semicond. Technol. Sci. **14**(5), 525–536 (2014)
18. M.S. Kim, W. Cane-Wissing, X. Li, J. Sampson, S. Datta, S.K. Gupta, V. Narayanan, Comparative area and parasitics analysis in FinFET and heterojunction vertical TFET standard cells. ACM J. Emerg. Technol. Comput. Syst. **12**(4) (2016). https://doi.org/10.1145/2914790
19. Nangate Library. http://www.nangate.com/. Accessed 26 Oct 2017
20. W. Xingsheng, A.R. Brown, B. Cheng, A. Asenov, Statistical variability and reliability in nanoscale FinFETs, in *IEEE International Electron Devices Meeting*, Dec 2011, pp. 5.4.1–5.4.4
21. E. Baravelli, M. Jurczak, N. Speciale, K. De Meyer, A. Dixit, Impact of LER and random dopant fluctuations on FinFET matching performance. IEEE Trans. Nanotechnol. **7**(3), 291–298 (2008)
22. J.-S. Yoon, C.-K. Baek, R.-H. Baek, Process-induced variations of 10-nm node bulk nFinFETs considering middle-of-line parasitics. IEEE Trans. Electron Devices **63**(9), 3399–3405 (2016)
23. Y.P. Tsividis, *Operation and Modeling of the MOS Transistor*. McGraw-Hill Series in Electrical Engineering (McGraw-Hill, New York, 1987)
24. R. Chau et al., High-k/metal-gate stack and its MOSFET characteristics. IEEE Electron Device Lett. **25**(6), 408–410 (2004)
25. M.T. Bohr, R.S. Chau, T. Ghani, K. Mistry, The high-k solution. IEEE Spectrum **44**(10), 29–35 (2005)
26. S.H. Rasouli, K. Endo, J.F. Chen, N. Singh, K. Banerjee, Grain-orientation induced quantum confinement variation in FinFETs and multi-gate ultra-thin body CMOS devices and implications for digital design. IEEE Trans. Electron Devices **58**(8), 2282–2292 (2011). https://doi.org/10.1109/TED.2011.2151196
27. H.F. Dadgour, K. Endo, V.K. De, K. Banerjee, Grain-orientation induced work function variation in nanoscale metal-gate transistors part I: modeling, analysis, and experimental validation. IEEE Trans. Electron Devices **57**(10), 2504–2514 (2010). https://doi.org/10.1109/TED.2010.2063191
28. W. Han, Z.M. Wang (eds.), *Toward Quantum FinFET*. Lecture Notes in Nanoscale 1 Science and Technology (Springer, New York, 2013). https://doi.org/10.1007/978-3-319-02021-1
29. C.A. Mack, *Field Guide to Optical Lithography* (SPIE Press, Bellingham, 2006)
30. C. Shin, *Variation-Aware Advanced CMOS Devices and SRAM* (Springer, Dordrecht, 2016)
31. E. Baravelli, A. Dixit, R. Rooyackers, M. Jurczak, N. Speciale, K. De Meyer, Impact of line-edge roughness on FinFET matching performance. IEEE Trans. Electron Devices **54**(9), 2466–2474 (2007). https://doi.org/10.1109/TED.2007.902166

32. V. Sriramkumar, N. Paydavosi, D. Lu, C.-H. Lin, M. Dunga, S. Yao, T. Morshed, A. Niknejad, C. Hu, BSIM-CMG 106.0. 0 Multi-Gate MOSFET Compact Model. Department of Electrical Engineering and Computer Sciences, University of California, Berkeley (2012)
33. K. Patel, T.-J.K. Liu, C.J. Spanos, Gate line edge roughness model for estimation of FinFET performance variability. IEEE Trans. Electron Devices **56**(12), 3055–3063 (2009). https://doi.org/10.1109/TED.2009.2032605
34. G. Leung, L. Lai, P. Gupta, C.O. Chui, Device- and circuit-level variability caused by line edge roughness for sub-32-nm FinFET technologies. IEEE Trans. Electron Devices **59**(8), 2057–2063 (2012). https://doi.org/10.1109/TED.2012.2199499
35. V.B. Kleeberger, H. Graeb, U. Schlichtmann, Predicting future product performance: modeling and evaluation of standard cells in FinFET technologies, in *Proceedings of the 50th Annual Design Automation Conference* (2013)
36. Y. Yang, N.K. Jha, FinPrin: FinFET logic circuit analysis and optimization under PVT variations. IEEE Trans. Very Large Scale Integr. Syst. **22**(12), 2462–2475 (2014)
37. S. Karapetyan, V.B. Kleeberger, U. Schlichtmann, FinFET-based product performance: modeling and evaluation of standard cells in FinFET technologies. Microelectron. Reliab. **61**, 30–34 (2016)
38. M. Alioto, G. Palumbo, M. Pennisi, Understanding the effect of process variations on the delay of static and domino logic. IEEE Trans. Very Large Scale Integr. Syst. **18**(5), 697–710 (2010)
39. S. Natarajan et al., A 14nm logic technology featuring 2nd-generation FinFET, air-gapped interconnects, self-aligned double patterning and a 0.0588 μm^2 SRAM cell size, in *IEEE International Electron Devices Meeting (IEDM)*, Dec 2014, pp. 3.7.1–3.7.3. https://doi.org/10.1109/IEDM.2014.7046976
40. Predictive Technology Model (PTM). ptm.asu.edu. Accessed 26 Oct 2017
41. J.P. Duarte, S. Khandelwal, A. Medury, C. Hu, P. Kushwaha, H. Agarwal, A. Dasgupta, Y.S. Chauhan, BSIM-CMG: standard FinFET compact model for advanced circuit design, in *41st European Solid-State Circuits Conference (ESSCIRC)*, Nov 2015
42. BSIM-CMG Model. http://bsim.berkeley.edu/models/bsimcmg/. Accessed 26 Oct 2017
43. ITRS 2012, ITRS. http://www.itrs2.net/. Accessed 26 Oct 2017
44. D.D. Lu, C.-H. Lin, A.M. Niknejad, C. Hu, Compact modeling of variation in FinFET SRAM cells. IEEE Des. Test Comput. **2**(27), 44–50 (2010)
45. X. Wang, B. Cheng, A.R. Brown, C. Millar, A. Asenov, Statistical variability in 14-nm node SOI FinFETs and its impact on corresponding 6T-SRAM cell design, in *Proceedings of the European Solid-State Device Research Conference (ESSDERC)* (2012)

Appendix A
Variance of a Function of Random Variables Approximated with Taylor's Theorem

This appendix illustrates the approximation of the mean and standard deviation of a function composed of several normal random variables by using a Taylor series expansion of first order.

Let us consider a function f composed of k normal random variables $(f(X_1, \ldots, X_k))$. Complexity to obtain the mean and standard deviation of this function can be lowered by approximating the function with a first-order Taylor series expansion and then applying some statistical properties to the resulting expression. The procedure to approximate a function with a Taylor series expansion can be found in Chap. 2. The variance of a function composed of k normal random variables $(f(X_1, \ldots, X_k))$ can be obtained with the following expression.

$$\mathrm{var}\left[f(X_1, \ldots, X_k)\right] = E\left[\left(f\left(X_1, \ldots, X_k\right) - f\left(\mu_{X_1}, \ldots, \mu_{X_k}\right)\right)^2\right] \tag{A.1}$$

For simplicity, a procedure to obtain the variance of a function of two random variables is illustrated, and then the results are extended to k random variables. Let us consider a function f of two normally distributed random variables X_1 $\left(\sim N\left(\mu_{X_1}, \sigma_{X_1}^2\right)\right)$ and X_2 $\left(\sim N\left(\mu_{X_2}, \sigma_{X_2}^2\right)\right)$,

$$f\left(X_1, X_2\right) \tag{A.2}$$

where $X_1 \sim N\left(\mu_{X_1}, \sigma_{X_1}^2\right)$ and $X_2 \sim N\left(\mu_{X_2}, \sigma_{X_2}^2\right)$ are random variables with normal distribution.

By expanding Eq. (A.2) with a *Taylor series* of first order around the nominal values of their variables and neglecting the *third and fourth central moments*, the following expression is obtained,

V. Champac, J. G. Gervacio, *Timing Performance of Nanometer Digital Circuits Under Process Variations*, Frontiers in Electronic Testing 39,
https://doi.org/10.1007/978-3-319-75465-9

$$f(X_1, X_2) = f(\mu_{X_1}, \mu_{X_2}) + \left[\frac{\partial f(X_1, X_2)}{\partial X_1}(X_1 - \mu_{X_1}) + \frac{\partial f(X_1, X_2)}{\partial X_2}(X_2 - \mu_{X_2})\right] \tag{A.3}$$

Moving the term $f(\mu_{X_1}, \mu_{X_2})$ to the left side in the previous equation, and squaring both the left and right sides of the resultant expression, the variance of f is obtained (Eq. (A.1)). This is,

$$\begin{aligned}\left(f(X_1, X_2) - f(\mu_{X_1}, \mu_{X_2})\right)^2 &= \left[\frac{\partial f(X_1, X_2)}{\partial X_1}(X_1 - \mu_{X_1}) + \frac{\partial f(X_1, X_2)}{\partial X_2}(X_2 - \mu_{X_2})\right]^2 \\ &= \left(\frac{\partial f(X_1, X_2)}{\partial X_1}\right)^2 (X_1 - \mu_{X_1})^2 + 2\frac{\partial f(X_1, X_2)}{\partial X_1}\frac{\partial f(X_1, X_2)}{\partial P_2} \\ &\quad \times (X_1 - \mu_{X_1})(X_2 - \mu_{X_2}) + \left(\frac{\partial f(X_1, X_2)}{\partial X_2}\right)^2 (X_2 - \mu_{X_2})^2\end{aligned} \tag{A.4}$$

By applying the *first moment* (expected value) to the left and right terms of (A.4), and using the property stating that the expected value of the sum of several terms is the sum of the expected value of each, an expression of the variance of the function f (σ_f^2) is obtained,

$$\begin{aligned}\sigma_f^2 = E&\left[\left(\frac{\partial f(X_1, X_2)}{\partial X_1}\right)^2 (X_1 - \mu_{X_1})^2\right] \\ +E&\left[2\frac{\partial f(X_1, X_2)}{\partial X_1}\frac{\partial f(X_1, X_2)}{\partial X_2}(X_1 - \mu_{X_1})(X_2 - \mu_{X_2})\right] \\ +E&\left[\left(\frac{\partial f(X_1, X_2)}{\partial X_2}\right)^2 (X_2 - \mu_{X_2})^2\right]\end{aligned} \tag{A.5}$$

The *linearity property* of the expected value is applied to each term of the previous equation,

$$\sigma_f^2 = \left(\frac{\partial f(X_1, X_2)}{\partial X_1}\right)^2 E\left[\left(X_1 - \mu_{X_1}\right)^2\right]$$
$$+2\frac{\partial f(X_1, X_2)}{\partial X_1}\frac{\partial f(X_1, X_2)}{\partial X_2}E\left[\left(X_1 - \mu_{X_1}\right)\left(X_2 - \mu_{X_2}\right)\right]$$
$$+\left(\frac{\partial f(X_1, X_2)}{\partial X_2}\right)^2 E\left[\left(X_2 - \mu_{X_2}\right)^2\right] \tag{A.6}$$

By replacing the *second central moment* (variance) in the previous equation,

$$\sigma_f^2 = \left(\frac{\partial f(X_1, X_2)}{\partial X_1}\right)^2 \sigma_{X_1}^2$$
$$+2\frac{\partial f(X_1, X_2)}{\partial X_1}\frac{\partial f(X_1, X_2)}{\partial X_2}E\left[\left(X_1 - \mu_{X_1}\right)\left(X_2 - \mu_{X_2}\right)\right]$$
$$+\left(\frac{\partial f(X_1, X_2)}{\partial X_2}\right)^2 \sigma_{X_2}^2 \tag{A.7}$$

By applying the definition of *covariance* of two random variables to the right side of the second term of the previous expression, the final expression of the variance of two random variables is obtained,

$$\sigma_f^2 = \left(\frac{\partial f(X_1, X_2)}{\partial X_1}\right)^2 \sigma_{X_1}^2$$
$$+2\frac{\partial f(X_1, X_2)}{\partial X_1}\frac{\partial f(X_1, X_2)}{\partial X_2}\rho_{X1,X2}\sigma_{X_1}\sigma_{X_2}$$
$$+\left(\frac{\partial f(X_1, X_2)}{\partial X_2}\right)^2 \sigma_{X_2}^2 \tag{A.8}$$

In (A.8), the partial derivatives represent the sensitivity of the function to small changes in the variables. Taking into account that $\mathrm{COV}(X_1, X_2) = \rho_{X1,X2}\sigma_{X_1}\sigma_{X_2}$, the previous equation can be rewritten as

$$\sigma_f^2 = S_{f,X_1}^2\sigma_{X_1}^2 + S_{f,X_2}^2\sigma_{X_2}^2 + 2S_{f,X_1}^2 S_{f,X_2}^2 \mathrm{COV}(X_1, X_2) \tag{A.9}$$

where S_{f,X_1}^2 and S_{f,X_2}^2 represent the sensitivity of the function to small changes of the parameters X_1 and X_2, respectively.

The result of the (A.8) for two random variables can be easily extended to k random variables. The variance of a function of k random variables $f(X_1, \ldots, X_k)$ is given by,

$$\sigma_f^2 = \sum_{i=1}^{n} \left(\frac{\partial f(X_1, \ldots, X_k)}{\partial X_i} \right)^2 \sigma_{X_i}^2$$

$$+ \sum_{i=1}^{n-1} \sum_{j=i+1}^{n} 2 \frac{\partial f(X_1, \ldots, X_k)}{\partial X_i} \frac{\partial f(X_1, \ldots, X_k)}{\partial X_j} \rho_{Xi,Xj} \sigma_{X_i} \sigma_{X_j} \quad (A.10)$$

and for independent random variables reduces to

$$\sigma_f^2 = \sum_{i=1}^{n} \left(\frac{\partial f(X_1, \ldots, X_k)}{\partial X_i} \right)^2 \sigma_{X_i}^2$$

Appendix B
Simulation of Process Variations with SPICE

This appendix describes the methodology to simulate the impact of process variations on the circuit performance using *Monte Carlo analysis* with the SPICE electrical simulator.

Monte Carlo analysis is a method that determines the statistical response of a circuit when process parameters are randomly varied according to a statistical distribution. *Monte Carlo analysis* allows calculating circuit variability to quantify the yield. *Monte Carlo analysis* in SPICE is run with another analysis such as DC, AC, or TRANsient.

Monte Carlo simulation repeats the circuit analysis a specific number of times; in each Monte Carlo run, a new value of the process parameter is randomly generated. It uses the following statement:

.<circuit analysis> tstep tstop sweep MONTE=val

where MONTE is the keyword to enable Monte Carlo analysis and *val* is the number of times that the circuit analysis is repeated. $tstep$ is the time step for the transient analysis, and $tstop$ is the time at which the transient simulation ends.

Example:

.TRAN 1ps 1ns sweep MONTE=100

The .PARAM statement is used to define a process parameter (L, W, T_{ox} and V_{th}) as a random variable according to a statistical distribution. The used *probability distribution* for the parameter is included to consider its statistical characteristics. This is defined as follows,

.PARAM < random parameter > = < probability distribution >

V. Champac, J. G. Gervacio, *Timing Performance of Nanometer Digital Circuits Under Process Variations*, Frontiers in Electronic Testing 39,
https://doi.org/10.1007/978-3-319-75465-9

Example:

.PARAM channel_length = GAUSS($nominal\_val, rel\_variation, number\_sigma$)
.PARAM channel_length = AGAUSS($nominal\_val, abs\_variation, number\_sigma$)

Both statements are used to define a process parameter as a random variable with a normal (Gaussian) distribution.

- The GAUSS parameter distribution function is used to generate a Gaussian distribution with a relative variation. The argument *nominal_val* represents the nominal value (*mean value*) of the parameter. *number_sigma* specifies the number of standard deviations to be considered to the right and the left of the parameter nominal value. *rel_variation* indicates the relative variation of the parameter, i.e., the percentage of deviation (divided by 100%) of the parameter about its nominal value.
- The AGAUSS parameter distribution function is used to generate a Gaussian distribution with an absolute variation. The arguments *nominal_val* and *number_sigma* represent the same parameters defined in the GAUSS function. *abs_variation* indicates the absolute variation of the parameter, i.e., the maximum deviation of the parameter in relation to its nominal value.

The *purely random* variation of process parameters are often defined using the GAUSS function. For instance, to assign independent threshold voltage values to each transistor, NMOS and PMOS, the following syntax can be used,

```
.PARAM
+VTN1=GAUSS(VTN, 0.25, 3)
+VTP1=GAUSS(VTP, 0.25, 3)
```

where VTN and VTP represent the nominal threshold voltage of the NMOS and PMOS transistor, respectively. With this statement, in each Monte Carlo simulation run, a different value is generated and assigned to each transistor by means of the VTN1 (NMOS) and VTP1 (PMOS) variables. In this instance, a relative variation of 25% representing 3σ of deviation about the nominal value is considered.

The *correlated* variation of process parameters are often defined using the AGAUSS function. For instance, to assign the same (correlated) channel length deviation to each transistor, NMOS and PMOS, the following syntax can be used,

```
.PARAM
+LNP=AGAUSS(0, AV, 3)
+LD=LNP
```

With this statement, in each Monte Carlo simulation run, the same channel length deviation can be assigned to all transistors of the cell by means of the variable LD. The generated value by the AGAUSS function is assigned to LNP; then this value is assigned to LD. LNP is used as an intermediate variable. In this instance, an absolute variation given by AV representing 3σ of deviation about the nominal value is considered.

The deviation of correlated process parameters, such as channel length (L), channel width (W), and oxide thickness (T_{ox}) can be included in the device declaration as follows:

$$L = L_{min} + LD \tag{B.1}$$

$$W = W_{min} + WD \tag{B.2}$$

$$T_{ox} = T_{oxnom} + T_{ox}D \tag{B.3}$$

where L_{min}, W_{min}, and T_{oxnom} represent the nominal values of the channel length, channel width, and oxide thickness, respectively. LD, WD, and $T_{ox}D$ represent the deviation of each parameter due to process variations, which are obtained using AGAUSS.

For a logic cell, the deviation of the correlated process parameters for its different transistors can be assumed to be the same because the distance among the transistors is very short. For a circuit with multiple gates, the correlation between logic cells must be determined using a spatial correlation model in order to obtain an accurate circuit delay (see Chap. 3 and Appendix C).

The .MEASURE statement is used to measure the signal delay in different points of the circuit respect a reference input. The following is an example of this statement:

```
.MEASURE TP TRIG V(IN) VAL=VDD/2 RISE=1 TARG V(OUT) VAL=VDD/2
  FALL=1
```

The use of this statement lets knows the mean and standard deviation of delay of a signal in the circuit.

Circuit Example

The impact of process variations on the delay of a two-input Nand gate (Nand-2) (see Fig. B.1) is simulated in SPICE. Purely random variations in the threshold voltage (V_{th}) and correlated variations in the channel length (L) are considered. The correlated process parameters for its different transistors of the Nand gate are assumed fully correlated because the separation distance among the transistors is very short. Table B.1 shows the transistor sizes of the gate.

Figure B.2 shows the SPICE netlist to simulate the impact of process variations in the threshold voltage (V_{th}) and channel length (L) on the delay of the Nand-2 gate using Monte Carlo analysis. The Monte Carlo analysis is declared with the transient analysis for 1000 runs. A higher number of Monte Carlo runs allow obtaining a

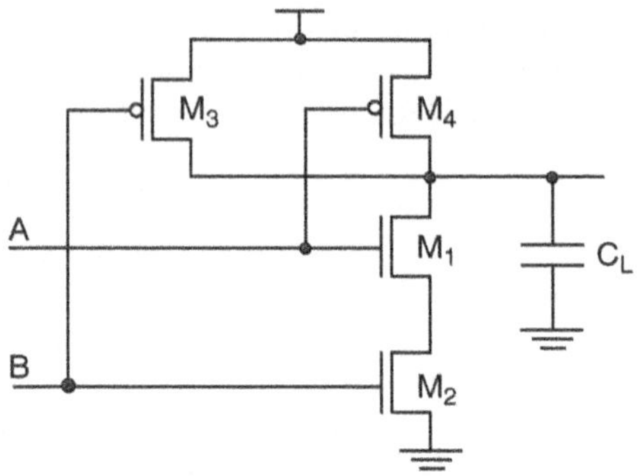

Fig. B.1 Two-input Nand gate

Table B.1 Size of the Nand-2 transistors

Transistor	Size (nm)
M_1	240
M_2	240
M_3	240
M_4	240

higher accuracy of the result. Values of the threshold voltage and channel length are generated using the GAUSS and AGAUSS functions, respectively, considering a 25% variation from the nominal value of V_{th} and 15% of variation about the nominal value of L. Threshold voltage deviation is added directly to the technology model file (tech.txt). Channel length deviation is added in the form of (B.1) on the device declaration.

Figure B.3 shows the SPICE Monte Carlo simulation results of the statistical delay performance of the Nand-2 gate. A load capacitance of $C = 2\,\text{fF}$ and a $0 \rightarrow 1$ transition at both inputs (A and B) are considered. Figure B.3a depicts the transient analysis response of the Nand-2 gate of the 1000 Monte Carlo iterations; observe that the output delay (OUT) of the gate presents multiple responses due to process variations in the threshold voltage and channel length. Figure B.3b illustrates the histogram of the transient analysis response. It shows that the statistical delay performance of the Nand-2 gate follows a Gaussian distribution, with a mean of 30.5 ps and standard deviation of 2 ps.

```
***** Monte Carlo analysis of a NAND-2 gate *****
***** TECHNOLOGY MODEL *****
.INCLUDE tech.txt

****** DEFINITION OF SUPPLY VOLTAGES AND INPUT SIGNALS *******
.global VDD
.global GND
VGN GND 0 DC 0
VDD VDD 0 DC 1.2
VVA A 0 PWL 0PS 0 100PS 0 110PS 1.2 1NS 1.2 ** 0- > 1 transition
VVB B 0 PWL 0PS 0 100PS 0 110PS 1.2 1NS 1.2 ** 0- > 1

****** NOMINAL VALUES OF PARAMETERS *******
.param
+Lmin=60e-3 ** minimum channel length
+Wmin=120e-3 ** minimum channel width
+VTN=0.4423 ** threshold voltage nmos
+VTP=-0.4308 ** threshold voltage pmos

******* GENERATION OF VALUES OF THE PARAMETERS ******
.param
+LNP=AGAUSS(0, 9e-3, 3)
+LD=LNP
+VTN1=GAUSS(VTN, 0.25, 3)
+VTP1=GAUSS(VTP, 0.25, 3)

********** DESCRIPTION OF THE NAND-2 GATE ************
M1 OUT A NX GND nmos w='2*Wmin' l='Lmin+LD'
M2 NX B GND GND nmos w='2*Wmin' l='Lmin+LD'
M3 OUT B VDD VDD pmos w='2*Wmin' l='Lmin+LD'
M4 OUT A VDD VDD pmos w='2*Wmin' l='Lmin+LD'

CL OUT 0 2fF ** load capacitance

****** DELAY MEASUREMENT ******
.MEASURE TPHL TRIG V(A) VAL=0.6 RISE=1 TARG V(OUT) VAL=0.6 FALL=1

****** MONTE CARLO ANALYSIS ******
.TRAN .1ps .3ns sweep MONTE=1000

.end
```

Fig. B.2 SPICE netlist to simulate the impact of threshold voltage (V_{th}) and channel length (L) variations on the Nand-2 gate delay using Monte Carlo analysis

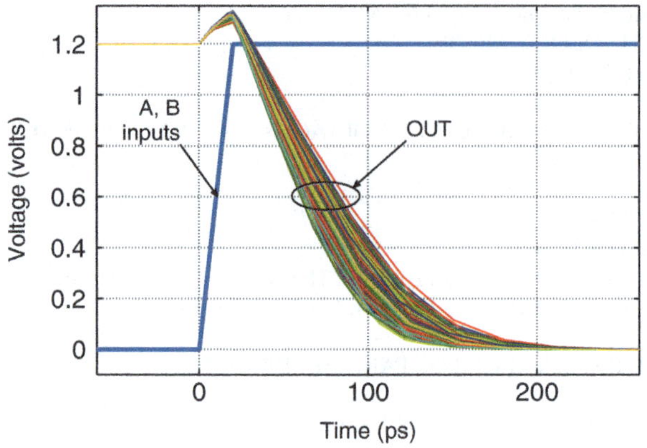

(a)

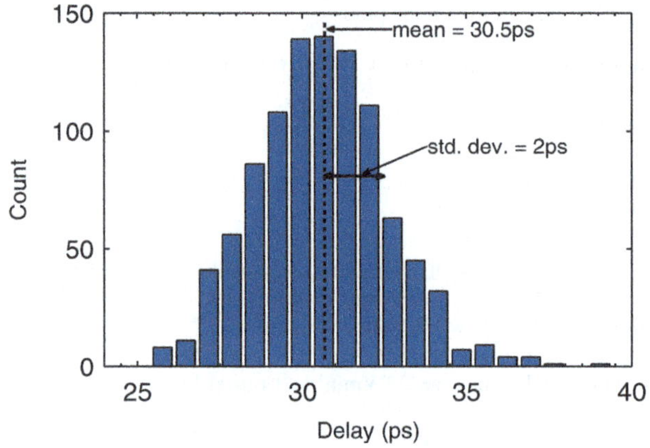

(b)

Fig. B.3 Statistical delay performance of a Nand-2 gate under process variations. (**a**) Transient response. (**b**) Histogram of delay

Appendix C
Simulation of Process Variations with SPICE Accounting Spatial Correlation

This appendix presents a SPICE-based approach to simulate the impact of process variations in the circuit performance, accounting for spatial correlation between different logic gates. As explained in Chap. 3, some process parameters (such as L, W and T_{ox}) exhibit spatial correlation due to manufacturing process variations consequently logic gates of a circuit are also spatially correlated. The spatial correlation has a significant impact on the standard deviation of the circuit delay; therefore, care should be taken for the proper modeling of this phenomenon using Monte Carlo analysis with SPICE.

As described in Appendix B, Monte Carlo analysis allows calculating circuit variability to quantify the yield. Monte Carlo simulation repeats the circuit analysis a specific number of times; in each Monte Carlo run, a new value of the process parameter is randomly generated. The Monte Carlo transient simulation is launched using the following statement,

```
.TRAN tstep tstop sweep MONTE=val
```

where $tstep$ is the time step for the transient analysis and $tstop$ is the time at which the transient simulation ends. val is the number of desired Monte Carlo sweeps.

Figure C.1 illustrates the process of performing Monte Carlo simulations of a logic path composed of three-inverter gates considering the spatial correlation among them. Considering that variation of the channel length of devices of a gate is correlated with the channel length of the other gates in the path. L_1, L_2, and L_3 represent the channel length of devices in the gate $G1$, $G2$, and $G3$, respectively. The circuit is simulated using a different set of channel length values for each gate. Monte Carlo simulation is executed val iterations. In the first Monte Carlo run, the channel lengths in each gate take different values ($L_1 = 68$ nm, $L_2 = 65$ nm,

V. Champac, J. G. Gervacio, *Timing Performance of Nanometer Digital Circuits Under Process Variations*, Frontiers in Electronic Testing 39,
https://doi.org/10.1007/978-3-319-75465-9

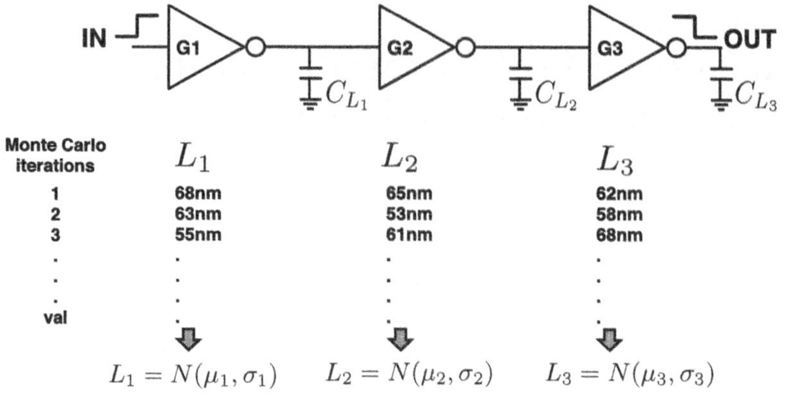

Fig. C.1 Inverter chain

and $L_3 = 62$ nm); in the second iteration, the channel lengths take other values ($L_1 = 63$ nm, $L_2 = 53$ nm, and $L_3 = 58$ nm); and so on until they complete the iterations (*val*) defined in the Monte Carlo statement. The way in which the parameter values change at each Monte Carlo sweep must satisfy the corresponding spatial correlation among gates, which is defined by the spatial correlation model (see Chap. 3).

It is important to say that values a parameter take must also satisfy the variability distribution of the parameter. These values can be obtained from the SPICE functions GAUSS or AGAUSS (see Appendix B). However, these functions generate uncorrelated random values of the parameters, which do not represent the real behavior of the process parameter due to spatial correlation. Therefore, to perform realistic Monte Carlo simulations of a circuit with SPICE, correlated parameters need to be generated via a mathematical treatment from the uncorrelated random values generated by the SPICE functions. A methodology to generate correlated values from uncorrelated ones is with the Cholesky transformation. The Cholesky transformation is commonly used in Monte Carlo analysis because it allows simulating circuits with multiple correlated variables. This is explained next.

C.1 Cholesky Transformation to Correlate Variables

A simple approach to generate a set of n correlated random variables ($Z_{c1}, Z_{c2}, \ldots, Z_{cn}$) from a set of n uncorrelated random variables ($Z_1, Z_2, \ldots, Z_n$) is to introduce a dependence relationship between them.

The *covariance* is a measure of the relationship between two random variables (see Chap. 2). The *correlation* can be understood as the normalized version of the covariance. In fact, the correlation is a measure of the linear relationship between

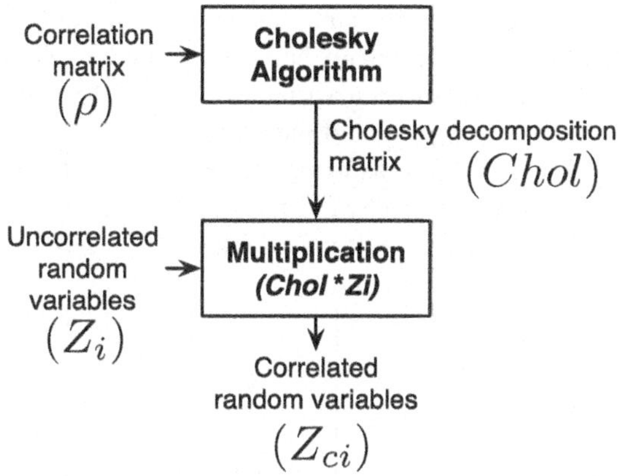

Fig. C.2 Cholesky transformation process to generate correlated random variables from uncorrelated ones

variables. Then, given a correlation matrix, it is possible to introduce a dependence relation between a set of variables.

The methodology that makes this possible is the Cholesky transformation. The Cholesky transformation is a method that generates correlated variables from uncorrelated ones and vice versa. Figure C.2 shows a flow diagram to generate correlated random variables using the Cholesky decomposition method. The inputs are the correlation matrix and the set of uncorrelated random variables. The output is the set of correlated random variables. Hence, given a correlation matrix (ρ), the Cholesky algorithm generates the Cholesky decomposition matrix (Chol). Then, Chol is multiplied by the uncorrelated random variables (Z_i) to obtain the set of correlated random variables (Z_{ci}).

Therefore, given a correlation matrix (ρ) of the form

$$\rho = \begin{pmatrix} r_{11} & r_{12} & \cdots & r_{1n} \\ r_{21} & r_{22} & \cdots & r_{2n} \\ \vdots & \vdots & \ddots & \vdots \\ r_{n1} & r_{n2} & \cdots & r_{nn} \end{pmatrix} \tag{C.1}$$

The correlation matrix can be factored as $(U^T * U)$, where U is an upper triangular matrix with positive diagonal values and U^T is its transpose. The matrix U is known as the Cholesky decomposition matrix.

The Cholesky decomposition matrix (Chol) is obtained after applying the Cholesky algorithm. The matrix has the form

$$\text{Chol} = \begin{pmatrix} c_{11} & c_{12} & \cdots & c_{1n} \\ c_{21} & c_{22} & \cdots & c_{2n} \\ \vdots & \vdots & \ddots & \vdots \\ c_{n1} & c_{n2} & \cdots & c_{nn} \end{pmatrix} \tag{C.2}$$

The coefficients c are the elements of the Cholesky decomposition (Chol) which define the relationship between variables. These coefficients carry the correlated information between variables. More information of the Cholesky decomposition can be found in related literature.

Finally, the transformation of uncorrelated random variables to correlated variables is performed by multiplying the Cholesky decomposition matrix with the array of uncorrelated random variables ($Z_{ci} = \text{Chol} * Z_i$) as follows:

$$\begin{bmatrix} Z_{c1} \\ Z_{c2} \\ \vdots \\ Z_{cn} \end{bmatrix} = \begin{bmatrix} c_{11} & c_{12} & \cdots & c_{1n} \\ c_{21} & c_{22} & \cdots & c_{2n} \\ \vdots & \vdots & \ddots & \vdots \\ c_{n1} & c_{n2} & \cdots & c_{nn} \end{bmatrix} \times \begin{bmatrix} Z_1 \\ Z_2 \\ \vdots \\ Z_n \end{bmatrix} \tag{C.3}$$

This transformation produces correlated random variables of the form

$$\begin{aligned} Z_{c1} &= c_{11} \cdot Z_1 + c_{12} \cdot Z_2 + \cdots + c_{1n} \cdot Z_n \\ Z_{c2} &= c_{21} \cdot Z_1 + c_{22} \cdot Z_2 + \cdots + c_{2n} \cdot Z_n \\ \vdots \quad & \quad \vdots \qquad\qquad \vdots \qquad\quad \ddots \qquad \vdots \\ Z_{cn} &= c_{n1} \cdot Z_1 + c_{n2} \cdot Z_2 + \cdots + c_{nn} \cdot Z_n \end{aligned} \tag{C.4}$$

The values of the Cholesky decomposition define the degree of correlation between the random variables.

Once (C.4) is obtained, it can be introduced in the SPICE netlist to generate correlated random variables for the Monte Carlo simulation. This process is illustrated in the next example.

C.2 Circuit Example

For the purpose of evaluating the statistical delay of the logic path (three-inverter chain) shown in Fig. C.1, and aiming for simplicity, only the spatial correlated variation on the channel length (L) is considered. Hence, three correlated random variables for the channel length (one for each inverter gate) need to be generated. The correlation values between each pair of gates due to their locations are computed using the exponential model described in Chap. 3; these are given in the next correlation matrix.

$$\rho = \begin{array}{c} \begin{matrix} G1 & \quad G2 & \quad G3 \end{matrix} \\ \begin{pmatrix} 1 & 0.812 & 0.624 \\ 0.812 & 1 & 0.743 \\ 0.624 & 0.743 & 1 \end{pmatrix} \end{array} \begin{matrix} G1 \\ G2 \\ G3 \end{matrix} \tag{C.5}$$

The Cholesky decomposition (C.6) of the correlation matrix (C.5) can be computed using the Cholesky algorithm. However, it should be mentioned that MATLAB tool incorporates the function *chol*(ρ), which automatically gives the Cholesky decomposition of a matrix.

$$\text{Chol} = \begin{pmatrix} 1.0000 & 0.8120 & 0.6240 \\ 0 & 0.5837 & 0.4049 \\ 0 & 0 & 0.6688 \end{pmatrix} \tag{C.6}$$

Once the Cholesky decomposition (C.6) is obtained, the expressions that model the correlated variation of the channel length between gates can be developed using (C.4). The three correlated random variables Z_{c1_L}, Z_{c2_L} and Z_{c3_L} representing the channel length of the $G1$, $G2$, and $G3$, respectively, are given by

$$\begin{aligned} Z_{c1_L} &= Z_{1_L} + 0.812 \cdot Z_{2_L} + 0.624 \cdot Z_{3_L} \\ Z_{c2_L} &= 0.5837 \cdot Z_{2_L} + 0.4049 \cdot Z_{3_L} \\ Z_{c3_L} &= 0.6688 \cdot Z_{3_L} \end{aligned} \tag{C.7}$$

where Z_{1_L}, Z_{2_L}, and Z_{3_L} represent the uncorrelated random variation of the channel length. Values of these parameters are generated using the SPICE function AGAUSS and the statistical characteristics of the channel length.

Equation (C.7) is included in the SPICE netlist to account for the correlated variation in the channel length between gates of the path. Figure C.3 shows the netlist used to simulate the three-inverter chain under process variations in the channel length, accounting for spatial correlation. At each Monte Carlo iteration, the statement

uncorrelated_l = AGAUSS(0, '60n*0.2', 3)

assigns a different and uncorrelated shift to parameters Z_{1_L}, Z_{2_L}, and Z_{3_L}. Then, values of these parameters are used in (C.7) to obtain the correlated parameters Z_{c1_L}, Z_{c2_L}, and Z_{c3_L}. Finally, the correlated variations are assigned to the corresponding channel length of the devices in each gate. In this example, a deviation of 20% from the nominal channel length (60 nm) and 1000 Monte Carlo sweeps are considered.

Figure C.4 shows the scatter plots between the measured propagation delays of the gates. It can be observed that propagation delays (t_{p_1} vs t_{p_2}) between gate $G1$ and $G2$ have a strong linear relationship (see Fig. C.4a); this behavior is because a high spatial correlation exists between them (see (C.5)). Propagation delays (t_{p_1} vs t_{p_3}) between gates $G1$ and $G3$ have a weaker relationship (see Fig. C.4b) than the one presented in gates $G1$ and $G2$; this behavior is due to a lower spatial correlation between them (see (C.5)). The relationship of propagation delays (t_{p_2} vs t_{p_3}) between gates $G2$ and $G3$ is shown in Fig. C.4c.

The correlation value between propagation delays of each pair of gates is given in Table C.1. These values are obtained using Pearson's correlation coefficient (see Chap. 2). Note that the obtained correlation between delays is similar to the

```
* Monte Carlo Simulations With Spatial Correlation
***** TECHNOLOGY MODEL *****
.INCLUDE 'tech.txt'

***** CORRELATED RANDOM VARIABLES GENERATION *****
.param uncorrelated_l = agauss(0, '60n*0.2', 3)
.param Z1L=uncorrelated_l
.param Z2L=uncorrelated_l
.param Z3L=uncorrelated_l
.param Zc1L='Z1L + 0.812*Z2L+ 0.624*Z3L'
.param Zc2L='0.5837*Z2L +0.4049*Z3L'
.param Zc3L='0.6688*Z3L'

************** DESCRIPTION OF THE CHAIN INVERTER PATH *****************
*INV1
Mp1 out1 in dd dd pmos w=240n l='60n+Zc1L'
Mn1 out1 in 0 0 nmos w=120n l='60n+Zc1L'
CL1 out1 0 4f
*INV2
Mp2 out2 out1 dd dd pmos w=240n l='60n+Zc2L'
Mn2 out2 out1 0 0 nmos w=120n l='60n+Zc2L'
CL2 out2 0 4f
*INV3
Mp3 out3 out2 dd dd pmos w=240n l='60n+Zc3L'
Mn3 out3 out2 0 0 nmos w=120n l='60n+Zc3L'
CL3 out3 0 4f
Vdd dd 0 1.2
Vin in 0 PULSE (0 1.2 10p 50p 50p 1n 2n )

*********** DELAY MEASUREMENTS *************
.MEASURE tran tp1 TRIG v(in) VAL=0.6 RISE=1 TARG v(out1) VAL=0.6 FALL=1
.MEASURE tran tp2 TRIG v(out1) VAL=0.6 FALL=1 TARG v(out2) VAL=0.6 RISE=1
.MEASURE tran tp3 TRIG v(out2) VAL=0.6 RISE=1 TARG v(out3) VAL=0.6 FALL=1
.MEASURE tran tppath TRIG v(in) VAL=0.6 RISE=1 TARG v(out3) VAL=0.6 FALL=1

*********** MONTE CARLO ANALYSIS ***************
.TRAN 1p 300p SWEEP MONTE=1000
.end
```

Fig. C.3 SPICE netlist to perform Monte Carlo simulations of a logic path, accounting for the spatial correlation between gates due to variations in the channel length

established spatial correlation between gates (see (C.5)), which proved that spatial correlation has properly accounted during simulations.

Monte Carlo simulations on SPICE provide the statistical information of delay performance of the circuit. Table C.2 gives the mean and standard deviation of delay of each gate and the path. The sum of each gate delay provides the total path delay.

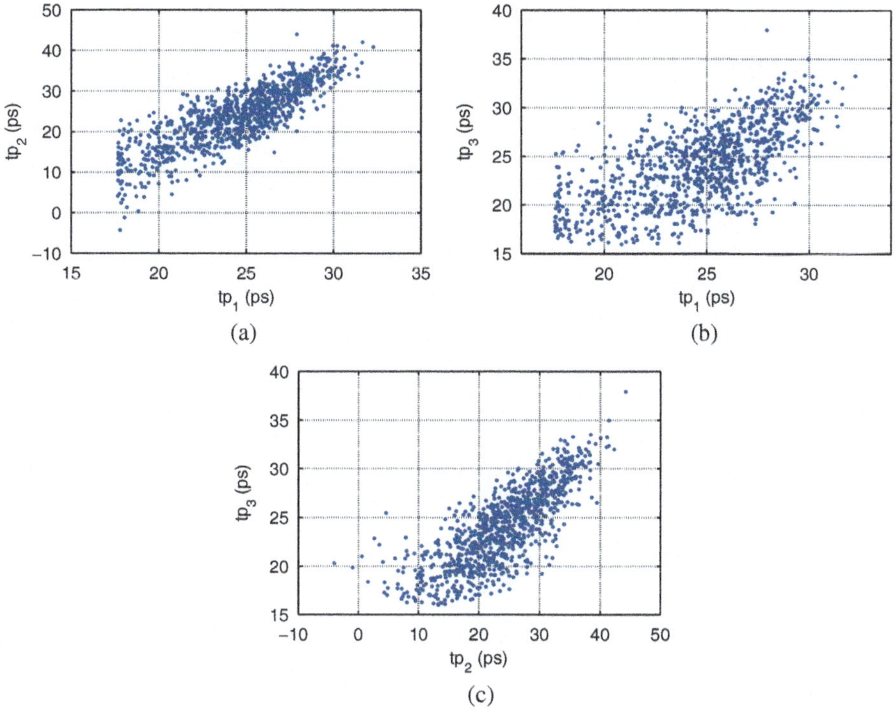

Fig. C.4 Scatter plots of the simulated path which show the relationship between the propagation delays of the gates. (**a**) t_{p_1} vs t_{p_2}. (**b**) t_{p_1} vs t_{p_3}. (**c**) t_{p_2} vs t_{p_3}

Table C.1 Correlation between measured propagation gate delays

Correlation	Value
$\rho_{G1,G2}$	0.838
$\rho_{G1,G3}$	0.661
$\rho_{G2,G3}$	0.798

Table C.2 Propagation delays of the three-inverter chain obtained with Monte Carlo simulation

Delay	Mean (ps)	Standard deviation (ps)
t_{p_1}	24.45	3.27
t_{p_2}	23.95	7.48
t_{p_3}	24.11	3.95
$t_{p_{\mathrm{path}}}$	72.46	13.80

Index

V. Champac, J. G. Gervacio, *Timing Performance of Nanometer Digital Circuits Under Process Variations*, Frontiers in Electronic Testing 39,
https://doi.org/10.1007/978-3-319-75465-9

The manufacturer's authorised representative in the EU is Springer Nature Customer Service Centre GmbH, Europaplatz 3, 69115 Heidelberg, Germany. If you have any concerns regarding our products, please contact ProductSafety@springernature.com

Printed and bound by CPI Group (UK) Ltd, Croydon, CR0 4YY
30/09/2026
02225643-0002